T4-AJX-497

T H E
UFO
FILES

JEROME CLARK
CONTRIBUTING WRITER

PUBLICATIONS INTERNATIONAL, LTD.

Contributing write JEROME CLARK is editor of
International UFO Reporter and columnist for *Fate*
magazine, which he edited for more than a decade. He is
author of the three-volume series *UFO Encyclopedia* and
coauthor of *New Age Encyclopedia.* Mr Clark is also vice
president of the Center for UFO Studies.

Consultant MARCELLO TRUZZI is an internationally
recognized scholar and consultant to many organizations,
including the Center for UFO Studies and *Strange*
magazine. He has appeared on *Nightline, Donahue, The
Oprah Winfrey Show, A Current Affair,* and *Geraldo.* Mr.
Truzzi is professor of sociology at Eastern Michigan
University and is the founder of the Center for Scientific
Anomalies Research.

Photo credits

Front cover: **M. Simpson/FPG International**

AP/Wide World Photos: 58, 61, 145, 158, 176, 183;
Bettmann Archives: 40, 51, 65, 80, 92, 133, 157, 159, 175,
181; **Center for UFO Studies:** 95; **Jerome Clark:** 76, 152;
Mary Evans Picture Library: 9, 63, 116, 119, 151; GEOS:
34, 82; **Fortean Picture Library:** 6, 12, 29, 33, 55, 122, 125,
147; Janet & Colin Bord: 17; PUFORG: 85; August C.
Roberts: 102; Dennis Stacy: 15, 107; Aaron Sussman: 25;
F.C. Taylor: 136; **Intercontinental U.F.O. Galactic
Spacecraft Research and Analytic Network Archives:**
165.

CONTENTS

BEYOND THE CLOUDS

Flying saucers, better known as unidentified flying objects or UFOs: Some people believe in their existence without question. The skepticism of others knows no bounds, and in fact the skeptics find much to ridicule in the world of ufology.

But the world is full of strange and eerie phenomena. To paraphrase Dr. J. Allen Hynek: Is ridicule the best approach to the phenomenon of UFOs? The scientific community has a duty and responsibility to give serious scientific examination to reports of UFO sightings. The first step toward an understanding of UFOs is to study the stories and photographs of those who have had some sort of contact with these phenomena.

The UFO Files presents the stories and case histories of individuals who claim to have seen or come in contact with UFOs. Some stories are patently false, while others are told by reputable people with supporting evidence. Some of the many photographs are obviously hoaxes, but other pictures taken by people who saw UFOs continue to defy conventional explanation, even decades after they were taken.

The first chapter, "In the Beginning," gives a quick overview of UFO history before Kenneth Arnold's sighting in 1947. The second chapter,

"UFOs: The Official Story," presents the official position of the United States government, which attempts to explain all UFO reports in terms of conventional phenomena. Chapter 3, "The Dimensions of a Phenomenon," explores the many ways UFOs have made their presence known on Earth: nocturnal lights; daylight discs; radar/visual detection; and close encounters of the first, second, and third kind. Chapter 4, "The Abduction Enigma," deals with stories of humans being abducted by aliens. Of course, UFOs provide fertile ground for hoaxes and scams of all sorts, and the fifth chapter, "Into the Wild Blue," details the fringe elements of ufology. The sixth chapter, "The Ultimate Secret," looks into the slow, painstaking unraveling of the most significant episode in ufology, the Roswell incident. Finally, the last chapter, "Beyond the Ultimate Secret," investigates claims that the U.S. government has developed a science-fictional aviation technology through a supersecret project.

Many possible explanations are offered for UFOs: natural phenomena, psychic visions, and, of course, visits by extraterrestrial beings. The idea of other life-forms somewhere in the universe stirs up an enormous amount of controversy. But such respected astronomers as Carl Sagan and Clyde Tombaugh believe the galaxy contains thousands of habitable planets, any number of which could support an advanced civilization.

So, turn these pages with an open mind and come explore the phenomenon of UFOs.

IN THE BEGINNING

Although strange aerial phenomena had been sighted for decades, it was Kenneth Arnold's report of "flying saucers" over Mount Rainier, Washington, on June 24, 1947, that brought unidentified flying objects into popular consciousness.

The date was June 24, 1947, a Tuesday; the time, just before 3:00 in the afternoon. Kenneth Arnold, a private pilot and fire-control-equipment salesman from Boise, Idaho, was flying over the Cascade Mountains searching for the remains of a lost C-46 for which a $5,000 reward had been offered.

Arnold never found the missing aircraft, but what he did see put his name in newspapers all over the world. He had just made a 180-degree turn over Mineral, Washington, when a bright flash of light startled him. During the next 30 seconds, Arnold frantically searched the sky for its source—he was afraid he was about to collide with another airplane. Then he saw another flash to his left, toward the north. When he looked in that direction, Arnold spotted nine objects, the lead one at a higher elevation than the rest, streaking south over Mount Baker toward Mount Rainier. Watching their progress from one peak to the next, he calculated their speed at 1,700 miles per hour. Even when he arbitrarily knocked 500 miles off that estimate, Arnold was still dealing with an impossible speed figure.

The objects, darting in and out of the smaller peaks, periodically flipped on their sides in unison. As they did so, the sunlight reflected off their lateral surfaces—thus explaining the flashes that had first caught his attention. Arnold wrote later, "They were flying diagonally in an echelon formation with a larger gap in their echelon between the first four and the last five." The lead object looked like a dark crescent; the other eight were

7

flat and disc-shaped. Arnold estimated that the chain they formed was five miles long. After two and a half minutes, they disappeared, heading south over Mount Adams. The age of unidentified flying objects (UFOs) had begun.

"FLYING SAUCERS"

The next day Arnold told his story to two reporters for Pendleton's *East Oregonian.* One of the reporters, Bill Bequette, put the story on the Associated Press wires. Within days, as similar sightings erupted around the country, an anonymous headline writer coined the phrase "flying saucers." But that name was not entirely original. On January 25, 1878, a Texas newspaper, the Denison *Daily News,* remarked on a local event that had taken place three days earlier. On the morning of January 22, farmer John Martin noted the swift passage through the southern sky of something like a "large saucer." The newspaper said, "Mr. Martin is a gentleman of undoubted veracity and this strange occurrence, if it was not a balloon, deserves the attention of our scientists."

The day Kenneth Arnold made his observations, as many as 18 other sightings of strange flying objects were reported in the Pacific Northwest. That morning prospector Fred M. Johnson had spotted five or six "round, metallic-looking discs" about 30 feet in diameter and 1,000 feet above him. He focused a telescope on one and saw that it had tails or fins (unlike those Arnold would observe a few hours later). For the

On the cover of its first issue, Fate *presented this imaginative version of Arnold's sighting, which was actually of distant objects. Inside, Arnold related his encounter with "flying saucers."*

duration of the sighting—close to a minute—Johnson's compass needle spun wildly, stopping only after the discs headed off to the southeast.

Actually, sightings of silvery discs had been going on since at least April 1947, when a U.S. Weather Bureau meteorologist and his staff had tracked a large, flat-bottomed ellipsoid as it shot from east to west over the skies of Richmond, Virginia. Sightings of similar objects took place the next month in Oklahoma, Colorado, Tennessee, Georgia, and New Jersey. These incidents went unnoted in the local press until after Kenneth Arnold's sighting opened the way to publication of such stories.

By the late 1940s Air Force investigators had taken to calling such things "unidentified flying objects." This was meant to be a neutral term, but skeptics complained that the words "flying" and "objects" implied both craft and intelligent guidance. Everyone could agree, though, that this phrase was better than the silly-sounding "flying saucers," which described only some of the aerial oddities people were reporting in the United States and around the world. Some of these phenomena looked like big metal cigars or fire-spewing torpedoes; others were spheres, triangles, or V shapes; and many were simply bright lights zigzagging across the night sky.

For the next half century, UFOs would be the focus of ceaseless controversy, wonderment, weirdness, fabrication, derision, mystification and, once in a while, serious investigation. Throughout this publication, many UFO phenomena are discussed; each story is presented from the perspective of the witness who experienced the event.

EARLY UFOS: FACT OR FAIRIES?

In A.D. 1211 Gervase of Tilbury, an English chronicler of historical events and curiosities, recorded this bizarre story:

> There happened in the borough of Cloera, one Sunday, while the people were at Mass, a marvel. In this town is a church dedicated to St. Kinarus. It befell that an anchor was dropped from the sky, with a rope attached to it, and one of the flukes caught in the arch above the church door. The people rushed out of the church and saw in the sky a ship with men on board, floating before the anchor cable, and they saw a man leap overboard and jump down to the anchor, as if to release it. He looked as if he were swimming in water. The folk rushed up and tried to seize him; but the Bishop forbade the people to hold the man, for it might kill him, he said. The man was freed, and hurried up to the ship, where the crew cut the rope and the ship sailed out of sight. But the anchor is in the church, and has been there ever since, as a testimony.

This tale—unrelated to any other British legend or supernatural tradition—is, according to

The appearance of black and white globes over Basel,
Switzerland, in 1566 was noted in a broadsheet.

folklorist Katharine Briggs, "one of those strange, unmotivated and therefore rather convincing tales that are scattered through the early chronicles."

In a 9th-century Latin manuscript, *Liber contra insulam vulgi opinionem,* the Archbishop of Lyons complained about the French peasantry's insistent belief in a "certain region called Magonia from whence come ships in the clouds." The occupants of these vessels "carry back to that region those fruits of the earth which are destroyed by hail and tempests; the sailors paying rewards to the storm wizards and themselves receiving corn and other produce." The archbishop said he had even witnessed the stoning to death of "three men and a woman who said they had fallen from these same ships." Jakob Grimm, a 19th-century folklorist, speculated, "'Magonia' takes us to some region where Latin was spoken, if we may rely on it referring to Magus, i.e., a magic land."

Are these early references to UFOs and aliens? Possibly. But references of this sort are few and far between. Although ancient and medieval records are filled with stories of strange shapes and figures in the sky, little in these accounts elicits visions of UFOs as we understand them today. Many eerie aerial phenomena of an earlier time can now be identified as meteors, comets, and auroral displays.

Still other accounts of UFOs are rooted in culture, perhaps the result of visions or hallucinations. Just before sunset on April 16, 1651, two women in rural England supposedly witnessed a battle between armies. At the conclusion of the battle there appeared, according to a contemporary account, blue angels "about the bigness of a capon, having faces (as they thought) like owls." Neither wars nor angels in the sky were uncommon "sights" from Roman times to the early modern era. In A.D. 793 the *Anglo-Saxon Chronicle* reported "fiery dragons . . . flying in the air," and almost a thousand years later, in 1762, a "twisting serpent" supposedly cavorted over Devonshire.

Along with this aerial activity were speculations and reports in popular lore of humanoid creatures dwelling in caves, bodies of water, or invisible realms. These humanoids varied widely in appearance; height alone ranged from a few inches to many feet. They possessed supernatural powers and sometimes kidnapped adults and children. These creatures, unpredictable and easily offended, were so feared that it was considered unwise to even speak their name. They were believed to be, according to one 17th-century account, "of a middle nature between man and angels." To see these humanoids, a person usually had to be in "faierie," meaning a state of enchantment. The traditional Anglo-Saxon name for these entities was "elves," now supplanted by "fairies."

Since 1947 some writers, notably Jacques Vallée in *Passport to Magonia,* have tried to link fairies to modern UFO encounters with humanoids. But this connection is speculative at best. The reader must be willing to assume that fairies were "real" and then overlook many dissimilarities between fairies and UFO humanoids. Fairy beliefs really have more in common with ghosts, monsters, and fabulous beasts than with modern accounts of encounters with UFOs.

Other writers, such as Desmond Leslie, George Hunt Williamson, M. K. Jessup, Yonah Fortner, and Brinsley le Poer Trench, also tried to find evidence of aliens visiting Earth before 1800, but their arguments are weak. Supposedly, extraterrestrials had been here for many thousands of years, leaving traces of their presence in legends and Biblical chapters as well as in such archaeological monuments as Stonehenge, the Great

Popular author Jacques Vallée rejects ufology's extraterrestrial hypothesis; his UFO solution emphasizes conspiracy theories and the occult.

Pyramid, and Peru's Nazca plains. These ideas were picked up and elaborated upon in the late 1960s and 1970s by a new school of writers (the most famous being Erich von Daniken of Switzerland), referring to "ancient astronauts."

Serious UFO researchers—not to mention astronomers, archaeologists, and historians—rejected these speculations, which in their view grew out of ignorance and distortion. Critics charged that no evidence supported so radical a revision of history and that such speculations deliberately slighted the role of human intelligence. Still, von Daniken's books had an enormous impact on impressionable readers.

THE ARRIVAL OF UFOS

In the 19th century, however, accounts of UFOs took on a more believable tone.

As day dawned June 1, 1853, students at Burritt College in Tennessee noticed two luminous, unusual objects just to the north of the rising sun. One looked like a "small new moon," the other a "large star." The first one slowly grew smaller until it was no longer visible, but the second grew larger and assumed a globular shape. (Probably the objects were moving in a direct line to and from the witnesses or were remaining stationary but altering their luminosity.) Professor A. C. Carnes, who interviewed the students and reported their sighting to *Scientific American,* wrote, "The first then became visible again, and increased rapidly in size, while the other diminished, and the two spots kept changing thus for about half an hour. There was considerable wind at the time, and light fleecy clouds passed by, showing the lights to be confined to one place."

Carnes speculated that "electricity" might be responsible for the phenomena. *Scientific American* believed this was "certainly" not the case; "possibly," the cause was "distant clouds of moisture." As explanations go, this was no more compelling than electricity. It would not be the last time a report and an explanation would make a poor match.

Unspectacular though it was, the event was certainly a UFO sighting, the type of sighting that could easily occur today. It represented a new phenomenon astronomers and lay observers were starting to notice with greater frequency in the earth's atmosphere. And some of these sights were startling indeed.

On July 13, 1860, a pale blue light engulfed the city of Wilmington, Delaware. Residents looked up into the evening sky to see its source: a 200-foot-long something streaking along on a level course 100 feet above. Trailing behind it at 100-foot intervals cruised three "very red and glowing balls." A fourth abruptly joined the other three after shooting out from the rear of the main object, which was "giving off sparkles after the manner of a rocket." The lead object turned toward the southeast, passed over the Delaware River, and then headed straight east until lost

Extraterrestrial lore claims England's Stonehenge has alien origins and supernatural powers.

from view. The incident—reported in the Wilmington *Tribune*, July 30, 1860—lasted for one minute.

During the 1850s and 1860s in Nebraska, settlers viewed some rather unnerving phenomena. Were they luminous "serpents"? Apparently not; instead they were elongated mechanical structures. A Nebraska folk ballad reported one such unusual sighting:

> 'Twas on a dark night in '66
> When we was layin' steel
> We seen a flyin' engine
> Without no wing or wheel
> It came a-roarin' in the sky
> With lights along the side
> And scales like a serpent's hide.

Something virtually identical was reported in a Chilean newspaper in April 1868 (and reprinted in *Zoologist*, July 1868). "On its body, elongated like a serpent," one of the alleged witnesses declared, "we could only see brilliant scales, which clashed together with a metallic sound as the strange animal turned its body in flight."

Lexicographer and linguist J.A.H. Murray was walking across the Oxford University campus on the evening of August 31, 1895, when he saw a

> brilliant luminous body which suddenly emerged over the tops of the trees before me on the left and moved eastward across the sky above and in front of me. Its

appearance was, at first glance, such as to suggest a brilliant meteor, considerably larger than Venus at her greatest brilliancy, but the slowness of the motion . . . made one doubt whether it was not some artificial firework. . . . I watched for a second or two till it neared its culminating point and was about to be hidden from me by the lofty College building, on which I sprang over the corner . . . and was enabled to see it through the space between the old and new buildings of the College, as it continued its course toward the eastern horizon. . . . [I]t became rapidly dimmer . . . and finally disappeared behind a tree. . . . The fact that it so perceptibly grew fainter as it receded seems to imply that it had not a very great elevation. . . . [I]ts course was slower than [that of] any meteor I have ever seen.

Some 20 minutes later, two other observers saw the same or a similar phenomenon, which they viewed as it traversed a "quarter of the heavens" during a five-minute period.

But in 1896 events turned up a notch: The world experienced its first great explosion of sightings of unidentified flying objects. The beginning of the UFO era can be dated from this year. Although sightings of UFOs had occurred in earlier decades, they were sporadic and apparently rare. Also, these earlier sightings did not come in the huge concentrations ("waves" in the lingo of ufologists, "flaps" to the U.S. Air Force) that characterize much of the UFO phenomenon between the 1890s and the 1990s.

Between the fall of 1896 and the spring of 1897, people began sighting "airships," first in California and then across most of the rest of the United States. Most people (though not all) thought the airships were machines built by secret inventors who would soon dazzle the world with a public announcement of a breakthrough in aviation technology leading to a heavier-than-air flying machine.

More than a few hoaxers were willing to play on this popular expectation, and sensation-seeking journalists were all too happy to join in. Newspaper stories quoted "witnesses" who claimed to have seen the airships land and to have communicated with the pilots. The pilots themselves were quoted word for word boasting of their aeronautical exploits and, in some instances, of their intention to drop "several tons of dynamite" on Spanish fortresses in Cuba. Any reader with access to more than one newspaper account could have seen that the stories conflicted wildly and were inherently unbelievable. We now know that no such ships existed in human technology, and no standard history of aviation ever mentions these tall tales.

But other sightings appear to have been quite real. Most descriptions were of a cylindrical object with a headlight, lights along the side, and a brilliant searchlight that swept the ground. Sometimes the objects were said to have huge wings. An "airship" was observed over Oakland, California, just after 8 P.M. on November 26. One witness said the object resembled "a great black cigar. . . . The body was at least 100 feet long and attached to it was a triangular tail, one apex being attached to the main body. The surface of the airship looked as if it were made of aluminum, which exposure to wind and weather had turned dark. . . . The airship went at tremendous speed" (Oakland *Tribune,* December 1, 1896). Witnesses in California numbered in the thousands, partly due to the objects' appearances—sometimes in broad daylight—over such major cities as Sacramento and San Francisco.

By February 1897 meandering nocturnal lights were also sighted in rural Nebraska. One of these lights swooped low over a group of worshippers leaving a prayer meeting: It turned out to be a cone-shaped structure with a headlight, three smaller lights along each side, and two wings. Such reports became the subject of newspaper articles around the state, leading the Kearney *Hub* on February 18 to remark that the "now famous California airship inventor is in our vicinity." In short order sightings were logged in Kansas, and by April the skies across a broad band of middle America—from the Dakotas and Texas in the west to Ohio and Tennessee in the east—were full of UFOs.

But the skies were also full of planets, stars, lighted balloons, and kites, which impressionable observers mistook for airships. Newspapers were full of outrageous yarns: A Martian perished in an airship crash in Texas. "Hideous" creatures lassoed a calf and flew off over Kansas with it. A "bellowing" giant broke the hip of a farmer who got too close to his airship after it landed in Michigan. These stories reflect a powerful undercurrent of speculation about extraterrestrial visitors.

The wave had run its course by May 1897, but cylindrical UFOs with searchlights would continue to be seen periodically for decades to come. A worldwide wave of sightings took place in 1909 in Australia, New Zealand, Great Britain, and the eastern United States. As late as 1957 an "airship" was seen over McMinnville, Oregon.

Witnesses reported other kinds of UFOs, too. One such report came from U.S. Navy Lieutenant Frank H. Schofield, who served as the Pacific Fleet's commander-in-chief in the 1930s. Standing on the deck of the USS *Supply* on February 28, 1904, Schofield and two other sailors watched "three remarkable meteors," bright red in color, as they flew beneath the clouds toward their ship. The objects then "appeared to soar, passing above the broken clouds . . . moving directly away from the earth. The largest had an apparent area of about six suns. It was egg-shaped, the larger end forward. The second was about twice the size of the sun, and the third, about the size of the sun. . . . The lights were in sight for over two minutes." (*Monthly Weather Review,* March 1904)

Far eerier stories lurked in the background. Only years later, when it was possible to talk about such things, did they come to light. One account surfaced more than 70 years later. In the summer of 1901, a 10-year-old Bournbrook, England, boy encountered something that looked like a box with a turret. Two little men clad in "military" uniforms and wearing caps with wires sticking out of them emerged through a door to wave him away. They then reentered the vehicle and flew away in a flash of light.

Similar events seem to have been occurring regularly over the early decades of the 20th century along with the less exotic sightings of strange aerial phenomena. These pre-1947 "close encounters of the third kind" were remarkably identical to the post-1947 reports in that the creatures who figured in the encounters were almost always reported to be human or humanoid in appearance. In Hamburg, Germany, in June 1914, several "dwarfs" about four feet tall were seen milling around a cigar-shaped vessel with lighted portholes; they then ran into the vessel and flew away. In Detroit during the summer of 1922, through windows along the perimeter of a hovering disc-shaped object, 20 bald-headed figures stared intently at a suitably bewildered young couple. At Christchurch, New Zealand, in August 1944, a nurse at a train station noticed an "upturned saucer" nearby. She approached it, looked through a rectangular window, and spotted two humanoid figures not quite four feet tall. A third figure stood just outside an open door. When this humanoid saw her, the being "drifted" through an open hatchway, and the "saucer" shot straight upward.

THE FIRST UFOLOGIST

Although these strange sky objects were reported with increasing frequency, the press and the scientific community treated each sighting as a one-time occurrence. There was no sense that such events, far from being isolated, were part of a larger phenomenon. Even the airship wave of 1896 and 1897 quickly passed out of the public's memory. But an eccentric American writer, Charles Fort (1874–1932), finally put it all together, becoming the world's first ufologist.

Born in Albany, New York, Fort was working as a newspaper reporter before age 20. Determined to become a writer, he traveled the world searching for experiences to write about. In South Africa Fort contracted a fever that followed him back to the United States. He married his nurse, Anna Filing, and embarked on a career as a freelance writer. Fort spent hours in the library pursuing his interests in nature and behavior. While paging through old newspapers and scientific journals, he began to notice, among other repeatedly chronicled oddities of the physical world, reports of strange aerial phenomena. Taking voluminous notes, he eventually turned out four books. The first three—*The Book of the Damned* (1919), *New Lands* (1923), and *Lo!* (1931)—dealt in part with UFO reports.

An intellectual with an impish sense of humor, Fort was fond of constructing outrageous "hypotheses" to "explain" his data. But beneath the humor Fort was trying to make a serious point: Scientists were refusing to acknowledge that the

Charles Fort, the first ufologist, wrote the first UFO book: The Book of the Damned, *published in 1919.*

world was full of weird phenomena and occurrences that did not fit with their theories. "Scientific" attempts to explain away such strange events as UFO sightings were laughably inadequate; their explanations, Fort wrote, were no less crazy than his own. "Science is established preposterousness," he declared. "Science of today—superstition of tomorrow. Science of tomorrow—superstition of today."

Behind the joking, however, Fort suspected that sightings of craftlike objects in the air indicated extraterrestrial visits to Earth. Yet he also understood humanity's resistance to such a fantastic, even threatening notion. In a letter published in the September 5, 1926, issue of *The New York Times,* Fort offered some prescient observations. Extraterrestrial beings would not have to hide their activities, he wrote, because if "it is not the conventional or respectable thing upon this earth to believe in visitors from other worlds, most of us could watch them a week and declare that they were something else, and likely enough make things disagreeable for anybody who thought otherwise."

CASE STUDIES

SAUCERS FROM WITHIN?

In the early 19th century an American eccentric, John Cleves Symmes (1779–1829), sought funding for an expedition to enter Earth through one of two 4,000-mile-wide polar holes. Inside Earth, he was convinced, a benevolent advanced civilization existed. Though an object of derision to most people, he was taken seriously by some, and the idea of a hollow Earth was championed in a number of books throughout the rest of the century and right into the next.

Today, hollow-Earthers believe flying saucers zip in and out of the polar holes. The people inside are descendants of Atlantis and its Pacific equivalent, Lemuria. There is even a strong Nazi wing of the movement. According to Canadian neo-Nazi Ernst Zundel, the principal advocate of this theory, Adolf Hitler and his elite troops escaped at the end of the second World War with their saucer technology into the hole at the South Pole.

CLOSE ENCOUNTER, 1897 STYLE

Late on the evening of April 13, 1897, as they were passing through Lake Elmo, Minnesota, on their way to Hudson, Wisconsin, Frederick Chamberlain and O. L. Jones spotted a shadowy figure in a clearing two blocks away. The figure carried a lantern and seemed to be looking for something. Thinking there might be some emergency, Chamberlain and Jones turned toward the clearing, but the figure and lantern disappeared into the trees. Moments later they heard the crackling of twigs and branches, followed by a "rushing noise . . . like the wind blowing around the eaves of a house," Chamberlain told the *St. Paul Pioneer Press* (April 15). "A second later and we distinguished a long, high object of a gray white color."

Although the two men could not get a clear view of it in the darkness, they could see that the object had two rows of red, green, and white lights on each side. It looked like "most of the top of a 'prairie schooner,'" Chamberlain said. It rose at a sharp angle, then headed south just above the treetops.

At the clearing, the two witnesses found impressed in the wet ground 14 prints, two feet long and six inches wide, arranged in an oblong pattern seven on a side. Apparently, these were traces left by the craft.

Furthermore, around that same time Adam Thielen, a farmer in the Lake Elmo area, heard a buzzing sound above him. When he looked up, he saw a dark object with red and green lights sailing overhead.

A VICTORIAN HUMANOID?

The attacker was tall and thin, had pointed ears and fiery eyes, and wore a cloak. He tore at his female victims' clothes and ripped their flesh with hands that felt like iron. When he escaped, he did not run; he bounced away. Witnesses who saw his feet swore that he had springs in his boot heels.

At first, the authorities had a hard time believing what victims were telling them. But by January 1838 so many Londoners had reported seeing the figure that the Lord Mayor formed a vigilance committee to capture "Springheel Jack."

In one especially notorious incident, he tried to snatch 18-year-old Jane Alsop right out of her own house. According to the London *Times* (February 22, 1838), he "presented a most hideous and frightful appearance, and vomited forth a quantity of blue and white flame from his mouth, and his eyes resembled red balls of fire. . . . [H]e wore a large helmet, and his dress, which appeared to fit him very tight, seemed to her to

An imaginative rendering of Springheel Jack at Aldershot, England, 1877.

resemble white oil skin." The young woman was saved by family members who literally pried her from Jack's clutches.

One day in 1845, in full view of frightened onlookers, Jack tossed a prostitute off a bridge; she drowned in the open sewer below. Sightings of a comparable figure were recorded elsewhere in England in 1877. In 1904 more than 100 residents of Everton saw a man in a flowing cloak and black boots making great leaps over streets and rooftops.

Who—or what—was Jack? Some suspect he was a rowdy nobleman, Henry, Marquis of Waterford, who died in 1859. Doubters countered that Jack-like leaps are physically impossible. German paratroopers during World War II who put springs in their boot heels got broken ankles for their efforts. Was Jack an extraterrestrial being? In July 1953, three Houston residents reported seeing a tall, bounding figure "wearing a black cape, skintight pants, and quarter-length boots." For a few minutes he remained visible in the pecan tree into which he had jumped. He disappeared shortly before a rocket-shaped UFO shot upward from across the street.

RUSTLERS FROM MARS

On April 23, 1897, a Kansas newspaper, the Yates Center *Farmer's Advocate,* reported an incredible story. On the evening of April 19, local rancher Alexander Hamilton, his son, and a hired man saw a giant cigar-shaped ship hovering above a corral near the house. Hamilton claimed that in a carriage underneath the structure were "six of the strangest beings I ever saw." Just then, the three men heard a calf bawling and found it trapped in the fence, a rope around its neck extending upward. "We tried to get it off but could not," Hamilton said, "so we cut the wire loose to see the ship, heifer and all, rise slowly, disappearing in the northwest."

The next day Hamilton went looking for the animal. He learned that a neighbor had found the butchered remains in his pasture. The neighbor, according to Hamilton, "was greatly mystified in not being able to find any tracks in the soft ground."

Hamilton's statement was followed by an affidavit signed by a dozen prominent citizens who swore that "for truth and veracity we have never heard [Hamilton's] word questioned." In the following days, his story was published in newspapers throughout the United States and even in Europe.

Ufologists rediscovered the account in the early 1960s, and the story rebounded to life through books and magazines. In 1976, however, an elderly Kansas woman came forward to say that shortly before the tale was reported in the *Farmer's Advocate,* she had heard Hamilton boast to his wife about the story he had made up. Hamilton belonged to a local liars' club that delighted in the concoction of outrageous tall tales. According to the woman, "The club soon broke up after the 'airship and cow' story. I guess that one had topped them all."

FOO FIGHTERS

A little-remembered cartoon character named Smokey Stover used to declare, "Where there's foo, there's fire." So when enigmatic aerial phenomena kept pace with airplanes and ships in both the European and Pacific theaters during World War II, someone called them "foo fighters." The name stuck. Nobody knew for sure what the foo fighters were, but it was usually assumed that the other side—either the Allies or the Axis powers—had developed a secret weapon. After the war's conclusion, it soon became clear that this was not the explanation.

With the arrival of "flying saucers" in the summer of 1947, memories of foo fighters were revived. Like UFOs after them, foo fighters came in assorted shapes and descriptions, from amorphous nocturnal lights—which gave them their name—to silvery discs.

A typical sighting of foos took place in December 1942 over France. A Royal Air Force pilot in a Hurricane interceptor saw two lights shooting from near the ground toward his 7,000-foot cruising altitude. At first he took the lights to be tracer fire. But when they ceased ascending and followed him, mimicking every evasive maneuver he made, the pilot realized they were under someone's intelligent control. The lights, which kept an even distance from each other all the while, pursued him for some miles.

In August of that same year, Marines in the Solomon Islands were startled to see a formation of 150 "roaring" silvery objects. Their color, one

This rare photograph of "foo fighters" shows UFOs of the World War II era. Reports of these objects were kept secret until 1944.

witness said, was "like highly polished silver." They had neither wings nor tails and moved (as later UFO witnesses would often remark) with a slight wobble.

Official censorship kept reports of these phenomena out of the newspapers until December 1944. All during the war, however, similar objects were sighted by both military and civilian observers in the United States.

UFOs: THE OFFICIAL STORY

*This sketch of the UFO sighted by Chiles and Whitted in 1948
(see page 36) is not quite accurate—witnesses reported two
rows of windows. Nevertheless, it shows an object of
structured appearance and extraordinary speed.*

From the beginning, society—in the persons of prominent scientists, government officials, military officers, journalists, and ordinary citizens— would make things disagreeable for those who insisted they had seen strange flying objects and those who believed them. Wherever there were "flying saucers," there was also ridicule, dished out in generous portions to anyone courageous or foolish enough to defy the reigning orthodoxy.

A 1951 *Cosmopolitan* article, prepared with Air Force cooperation and encouragement, lashed out at the "screwballs" and "true believers" who thought they were seeing flying saucers. In the decades to come, others would accuse UFO observers of every conceivable social crime or mental disorder. As a result, only a small minority of witnesses would ever report their sightings, and many who did soon lived to regret it. In 1977 a group of professional debunkers warned *The New York Times* that belief in UFOs is not only irrational but also dangerous; if sufficiently widespread, civilization itself could collapse.

Yet in the face of jeering derision and inflated rhetoric, the sightings continued. The great majority of sightings would be by individuals who would have been implicitly believed had they been testifying to anything less outrageous. Of course, these witnesses were not always right. Even sympathetic investigators found that most reports could be explained conventionally. Few of the reports were outright hoaxes (around one percent, according to the Air Force's estimate), but sane and sober eyewitnesses often mistook weather balloons, stars and planets, advertising planes, and other ordinary objects for extraordinary objects. Still, some sightings stubbornly resisted explanation.

In the summer of 1947, the Air Materiel Command (AMC) was asked to study the situation and make recommendations about what should be done. On September 23 Lieutenant General Nathan F. Twining, the AMC head, wrote his superior with this analysis: "The phenomenon reported is something real and not visionary or fictitious." Three months later the Air Force established Project Sign under AMC command, headquartered at Wright Field, which was soon to be Wright-Patterson Air Force Base (AFB), Dayton, Ohio, to investigate UFO reports.

AN "ESTIMATE OF THE SITUATION"

By late July 1948, Project Sign investigators had come to an incredible conclusion: Visitors from outer space had arrived. They had begun with suspicions. Now they had the proof. The proof was . . . well, it depends on which of two versions of the story is to be believed.

In the better-known version, the proof arrived in the sky southwest of Montgomery, Alabama, at 2:45 A.M. on July 24, 1948. To Clarence S. Chiles and John B. Whitted, pilot and copilot of an Eastern Airlines DC-3, the object at first looked like a distant jet aircraft to their right and just above them. But it was moving awfully fast. Seconds later, as it streaked past them, they saw something that Whitted thought looked like "one of those fantastic Flash Gordon rocket ships in the funny papers." It was a huge, tube-shaped structure, its fuselage three times the circumference of a B-29 bomber, and with two rows of square windows emanating white light. It was, Chiles would remember, "powered by some jet or other type of power shooting flame from the rear some 50 feet." The object was also glimpsed by the one passenger who was not sleeping. After it passed the DC-3, it shot up 500 feet and was lost in the clouds at 6,000 feet altitude.

Although Chiles and Whitted didn't know it at the time, an hour earlier a ground maintenance crewman at Robins AFB, Georgia, had seen the same or an identical object. On July 20, observers in The Hague, the Netherlands, had watched a comparable craft move swiftly through the clouds.

It took investigators little time to establish that no earthly missile or aircraft could have been responsible for these sightings. Moreover, independent verification of the object's appearance and performance left no question of the witnesses' being mistaken about what they had seen. In the days following the sighting, Project Sign prepared an "estimate of the situation"—a thick document stamped TOP SECRET—that argued that this and other reliably observed UFOs could only be otherworldly vehicles. But when the estimate landed on the desk of Air Force Chief of Staff General Hoyt S. Vandenberg, he promptly rejected it on the grounds that the report had not proved its case.

In short order Project Sign's advocates of extraterrestrial visitation were reassigned or encouraged to leave the service. The Air Force then embarked on a debunking campaign interrupted only for the brief period between 1951 and 1953 when Captain Edward J. Ruppelt, who took an open-minded approach, headed the official UFO project. Project Sign was succeeded by Project Grudge (1949–1952); Project Blue Book, established in March 1952, succeeded Project Grudge. Practically until the day the Air Force closed down Project Blue Book in December 1969, it denied that such a document had ever existed, even when former UFO-project officers swore they had seen or heard of it. No one could produce a copy of the document, however, because the Air Force had ordered all copies burned.

At least one source disputes this account, on the authority of Captain Ruppelt, who tells it in his memoir of his Project Blue Book years, *The Report on Unidentified Flying Objects* (1956). Years after the original incidents, a retired AMC-assigned officer (now deceased) claimed that Project Sign prepared two drafts of the estimate. The first draft referred to what the officer remembered as a "physical evidence" case in New Mexico. When Vandenberg saw this reference, he demanded its removal. The second draft, with the offending paragraphs deleted, argued its case solely from eyewitness testimony—of which the Chiles-Whitted encounter was an impressive example. Vandenberg could now claim that in the absence of physical evidence, no proof existed.

A long time would pass before civilian investigators learned of this New Mexico physical-evidence case. It would turn out to be one of the most important incidents—perhaps the most important incident—in UFO history. With these revelations would come the belated realization that ufology has two histories: a public one and a hidden one. But we're getting ahead of ourselves.

THE INVADERS

Flying saucers were supposed to be a fad. Pundits tied these strange shapes in the sky to "war nerves," a sort of delayed-response reaction to the traumas of World War II. They were also supposed to be a peculiarly American delusion. Unidentified flying objects, however, have survived longer than war memories and remain an eerie, discomforting presence throughout the world.

As military and civilian researchers scrambled to make sense of all this, anything seemed possible—even attack by hostile aliens.

On January 7, 1948, Kentucky Air National Guard Captain Thomas F. Mantell, Jr., died when his F-51 crashed after chasing what he called, in one of his last radio transmissions, "a metallic object of tremendous size." The official Air Force line was that Mantell saw Venus. Unofficially, many officers feared that a spaceship had shot down Mantell's plane with a frighteningly superior extraterrestrial weapon.

Neither answer, it turned out, was correct. Declassified documents eventually disclosed that

National Guard pilot Thomas F. Mantell, Jr.

the Navy had been conducting secret balloon experiments as part of its Skyhook project, which sought to measure radiation levels in the upper atmosphere. As Mantell pursued what he apparently thought was a spaceship, he had foolishly ascended to 25,000 feet. This was a dangerous altitude for the aircraft he was piloting, and Mantell blacked out from lack of oxygen. His F-51 spun out of control and crash-landed in the front lawn of a farmhouse near Franklin, Kentucky. But in the days that followed the tragedy, sensational headlines fueled everyone's worst fears about flying saucers, and the Mantell incident entered UFO legend.

Just as frightening, though more bizarre and less publicized, was a fatal encounter that occurred on November 23, 1953, over Lake Superior. That evening, as Air Defense Command radar tracked an unidentified target moving at 500 miles per hour over the lake, an F-89C all-weather jet interceptor from Kinross AFB took off in hot pursuit. Radar operators watched the aircraft close in on the UFO, and then something fantastic happened: The two blips merged and then faded on the screen, and all communication with the interceptor ceased. An extensive land and water search found not a trace of the craft or the two men aboard it: pilot Lieutenant Felix Moncla, Jr., and radar observer Lieutenant R. R. Wilson.

Unlike the Mantell incident, the Kinross case attracted minimal newspaper coverage. Also unlike the Mantell case, the Kinross case has never been satisfactorily explained. Later, after aviation writer Donald E. Keyhoe broke the story in his best-selling book *The Flying Saucer Conspiracy* (1955), the Air Force insisted that the "UFO" had proved on investigation to be a Royal Canadian Air Force C-47. The F-89C had not actually collided with the Canadian transport plane, but something unspecified had happened, and the interceptor crashed. Aside from implying woeful incompetence on the radar operators' part, this "explanation"—still the official one—flies in the face of the Canadian government's repeated denials that any such incident involving one of its aircraft ever took place.

In 1958 Keyhoe got hold of a leaked Air Force document that made it clear that officialdom considered the Kinross incident a UFO encounter of the strangest kind. The document quoted these words from a radar observer who had been there: "It seems incredible, but the blip apparently just swallowed our F-89." The following year, in conversations with civilian ufologists Tom Comella and Edgar Smith, Master Sergeant O. D. Hill of Project Blue Book confided that such incidents—he claimed Kinross had not been the only one—had officials worried. Many, he said, believed UFOs to be of extraterrestrial origin and wanted to prevent an interplanetary Pearl Harbor. Comella subsequently confronted Hill's superior, Captain George T. Gregory, at Blue Book headquarters. Gregory looked shocked, left the room for a short period, and returned to state, "Well, we just cannot talk about those cases."

THE INTELLIGENCE COMMUNITY TAKES CONTROL

A few minutes before midnight on Saturday, July 19, 1952, an air traffic controller at National Airport in Washington, D.C., noticed some odd blips on his radar screen. Knowing that no aircraft were flying in that area—15 miles to the southwest of the capital—he rushed to inform his boss, Harry G. Barnes. Barnes recalled a few days later, "We knew immediately that a very strange situation existed. . . . [T]heir movements were completely radical compared to those of ordinary aircraft." They moved with such sudden bursts of intense speed that radar could not track them continuously.

Soon, National Airport's other radar, Tower Central (set on short-range detection, unlike Barnes's Airway Traffic Control Central [ARTC]), was tracking unknowns. At Andrews AFB, ten miles to the east, Air Force personnel gaped incredulously as bright orange objects in the southern sky circled, stopped abruptly, and then streaked off at blinding speeds. Radar at Andrews AFB also picked up the strange phenomenon.

The sightings and radar trackings continued until 3 A.M. By then witnesses on the ground and in the air had observed the UFOs, and at times all three radar sets had tracked them simultaneously.

Exciting and scary as all this had been, it was just the beginning of an incredible episode. The next evening radar tracked UFOs as they performed extraordinary "gyrations and reversals," in the words of one Air Force weather observer. Moving at more than 900 miles per hour, the objects gave off radar echoes exactly like those of aircraft or other solid targets. Sightings and trackings occurred intermittently during the week and then erupted into a frenzy over the following weekend. At one point, as an F-94 moved on targets ten miles away, the UFOs turned and darted en masse toward the interceptor, surrounding it in seconds. The badly shaken pilot, Lieutenant William Patterson, radioed Andrews AFB to ask if he should open fire. The answer, according to Albert M. Chop, a civilian working as a press spokesperson for the Air Force who was present, was "stunned silence. . . . After a tense moment, the UFOs pulled away and left the scene."

As papers, politicians, and public clamored for answers, the Air Force hosted the biggest press conference in history. A transcript shows that the spokesperson engaged in what amounted to double-talk, but the reporters, desperate for something to show their editors, picked up on Captain Roy James's off-the-cuff suggestion that temperature inversions had caused the radar blips. James, a UFO skeptic, had arrived in Washington only that morning and had not participated in the ongoing investigation.

Nonetheless, headlines across the country echoed the sentiments expressed in the Washington *Daily News:* "SAUCER" ALARM DISCOUNTED BY PENTAGON; RADAR OBJECTS LAID TO COLD AIR FORMATIONS. This "explanation" got absolutely no support from those who had seen the objects either in the air or on the radar screens, and the U.S. Weather Bureau, in a little-noted statement, rejected the theory. In fact, the official Air Force position, which it had success-fully obscured, was that the objects were "unknowns."

But while the nation's opinion makers, satis-fied that all was well, went on to other stories, the aftershocks of the Washington UFO invasion reverberated throughout the defense establish-ment. H. Marshall Chadwell, assistant director of the CIA's Office of Scientific Intelligence, warned CIA director General Walter Bedell Smith, "At any moment of attack [from the Soviet Union], we are now in a position where we cannot, on an instant basis, distinguish hardware from phantom, and as tension mounts we will run the increasing risk of false alerts and the even greater danger of false-ly identifying the real as phantom." Chadwell feared that the Soviets could plant UFO stories as a psychological warfare exercise to sow "mass hysteria and panic." In fact, as *The New York Times* noted in an August 1, 1952, analysis, the Washington sightings and others across the coun-try in July were so numerous that "regular intel-ligence work had been affected."

In fact, during the Washington events, traffic related to the UFO sightings had clogged all intel-ligence channels. If the Soviets had chosen to take advantage of the resulting paralysis to launch an air or ground invasion of the United States, the appropriate warnings would have had no way to get through.

Determined that this would never happen again, the CIA approached Project Blue Book and said it wanted to review the UFO data accumulated since 1947. In mid-January 1953 a scientific panel headed by CIA physicist H. P. Robertson briefly reviewed the Air Force material, dismissed it quickly, and went on to its real business: recommending ways American citizens could be discouraged from seeing, reporting, or believing in flying saucers. The Air Force should initiate a "debunking" campaign and enlist the services of celebrities on the unreality of UFOs. Beyond that official police agencies should monitor civilian UFO research groups "because of their potentially great influence on mass thinking. . . . The apparent irresponsibility and the possible use of such groups for subversive purposes should be kept in mind."

The panel's existence and its conclusions remained secret for years, but the impact on official UFO policy was enormous. In short order Project Blue Book was downgraded, becoming little more than a public-relations exercise. In 1966 the Air Force sponsored a project, directed by University of Colorado physicist Edward U. Condon, to conduct what was billed as an "independent" study. In fact it was part of an elaborate scheme to allow the Air Force, publicly anyway, to get out of the UFO business.

The Condon committee was to review or reinvestigate Project Blue Book data and decide if further investigation was warranted. As an internal memorandum leaked to *Look* magazine in 1968 showed, Condon and his chief assistant knew before they started that they were to reach negative conclusions. Condon sparked a firestorm of controversy when he summarily dismissed two investigators who, not having gotten the message, returned from the field with positive findings. In January 1969, when the committee's final report was released in book form, readers who did not get past Condon's introduction were led to believe that "further extensive study of UFOs probably cannot be justified on the expectation that science will be advanced thereby." Those who bothered to read the book found that fully one-third of the cases examined remained unexplained, and scientist-critics would later note that even some of the "explained" reports were unconvincingly accounted for.

But that did not matter; Condon and his committee had done their job, and the Air Force closed down Project Blue Book at the end of the year.

Some years later a revealing memo came to light through the Freedom of Information Act. It amounted to confirmation of a long-standing suspicion: Project Blue Book served as a front for a classified project that handled the truly sensitive reports. The memo, prepared on October 20, 1969, by Brigadier General C. H. Bolender, the Air Force's Deputy Director of Development, noted that "reports of UFOs which could affect national security should continue to be handled through the standard Air Force procedure designed for this purpose." He did not explain what this "standard Air Force procedure" was, and the 16 pages attached to his memo—which presumably would have shed some light on this curious assertion—are missing from the Air Force files.

The Bolender memo was the first whiff from the cover-up's smoking gun. There would be more—a lot more—in the years to come.

CASE STUDIES

BUBBLES FROM OUTER SPACE

Not all unidentified flying objects are potential alien spacecraft. Some are much stranger. Consider, for example, the strange phenomenon that passed over Biskopsberga, Sweden, early in the 19th century.

It was a cloudless afternoon on May 16, 1808, with a hard wind blowing in from the west; the sun over the village suddenly grew dim. At the western horizon a great number of spherical bodies appeared. They were heading toward the sun and changed from dark brown to black as they got closer to the sun. As they approached, they lost speed but sped up again after passing in front of the sun. They moved in a straight procession across the sky to the eastern horizon. According to *Transactions of the Swedish Academy of Sciences* (1808), "The phenomenon lasted uninterruptedly, upwards of two hours, during which time millions of similar bodies continually rose in the west, one after the other irregularly, and continued their career in exactly the same manner."

Some of the balls fell out of the sky, several landing not far from K. G. Wettermark, secretary of the Swedish Academy of Sciences. Seen just before they hit the ground, they resembled "those air-bubbles which children use to produce from soapsuds by means of a reed. When the spot, where such a ball had fallen, was immediately after examined, nothing was to be seen, but a scarcely perceptible film or pellicle, as thin and fine as a cobweb, which was still changing colors, but soon entirely dried up and vanished."

The balls still in the air continued their passage until all disappeared in the east.

ELINT vs. UFO

Possessing the most sophisticated electronic intelligence (ELINT) gear available to the U.S. Air Force, the RB-47 could handle anything. Unfortunately, in the morning hours of July 17, 1957, over the southern United States, an RB-47 came across something it was unprepared for.

In the first hint of what was to come, one of the three officers who operate the electronic countermeasures (ECM) equipment detected an odd signal. Moving up the radar screen, the blip passed some distance in front of the RB-47, then over Mississippi. Though puzzled, the officer said nothing. However, a few minutes later, at 4:10 A.M., the sudden appearance of an intense blue light bearing down on the aircraft shook the pilot and copilot. Even more unnerving, the object changed course in the blink of an eye and disappeared at the two o'clock position. The aircraft radar picked up a strong signal in the same spot. The UFO maintained this position even as the RB-47 continued toward east Texas.

The pilot then observed a "huge" light, attached, he suspected, to an even bigger something that the darkness obscured. When the electronics gear noted the presence of another UFO in the same general location as the first, the pilot turned the plane and accelerated toward it. The UFO shot away. By now the crew had alerted the Duncanville, Texas, Air Force ground radar station, and it was soon tracking the one UFO that remained (the second had disappeared after a brief time). At 4:50 radar showed the UFO abruptly stopping as the RB-47 passed under it. Barely seconds later it was gone.

This incredible case—considered one of the most significant UFO encounters ever—remained classified for years. When it became known years later, the Air Force declared that the RB-47 crew had tracked an airliner. Physicist Gordon David Thayer, who investigated the incident for the University of Colorado UFO Project, called this explanation "literally ridiculous."

A ROARING UFO AND FIGURES AT SOCORRO

Officer Lonnie Zamora was chasing a speeder south of Socorro, New Mexico, late on the afternoon of April 24, 1964, but he was about to enter UFO history. No less than the head of Project Blue Book would later tell a CIA audience that Za-

Zamora (left) with Air Force investigators from Project Blue Book.

mora's experience was the most puzzling UFO case he had ever dealt with.

All Zamora knew at first was that a roar was filling his ears and a flame was descending in the southwestern sky. Breaking off the chase, Zamora sped to the site, where he expected to find that a dynamite shack had exploded. Instead, as he maneuvered through the hilly terrain, he glimpsed a shiny, car-size object resting on the ground about 150 yards away. Near it stood two small figures clothed in what looked like white coveralls.

Zamora briefly lost sight of the object and the figures as he passed behind a hill. Zamora thought he had witnessed a car accident, but when he got out of his car to investigate, he suddenly realized otherwise. Egg-shaped and standing on four legs, the object displayed a peculiar

51

insignia on its side, something like an arrow pointing vertically from a horizontal base to a half-circle crown. The two figures had disappeared, and the object was emitting an ominous roar again. Frightened, Zamora charged back to his car. At one point he glanced over his shoulder to see the UFO, now airborne, heading toward a nearby canyon.

Project Blue Book investigators found that Zamora had a reputation for integrity. They also examined what looked like landing marks on the desert floor. In the middle of these marks was a burned area, apparently from the craft's exhaust.

KEYHOE PRESSES THE AIR FORCE

In the 1950s the Air Force's most forceful critic, retired Marine Corps Major Donald E. Keyhoe, caused Pentagon UFO debunkers no end of consternation. A respected aviation journalist, Keyhoe wrote an explosive article, "The Flying Saucers Are Real," for the widely read men's magazine *True* (January 1950 issue). Not only did intelligent beings from elsewhere have Earth under scrutiny, Keyhoe claimed, but the Air Force knew it and was conspiring to cover up the truth. From his Washington contacts Keyhoe gathered leaked information about encounters between military interceptor aircraft and fast-moving discs as well as documents suggesting concern about these events.

In 1957, after writing three best-selling books on the UFO cover-up, Keyhoe became director of the National Investigations Committee on Aerial

Phenomena (NICAP). Though Keyhoe had powerful allies, including former CIA chief R. H. Hillenkoetter, the Air Force had him outgunned. Keyhoe retired from the fray in 1969. When he died 19 years later, the battle against official secrecy had passed to other, younger hands. His pioneering efforts are still remembered.

THE SENATOR'S SOVIET SAUCERS

Georgia Senator Richard Russell was a major figure in the U.S. Senate. As head of the Senate Armed Services Committee, he exerted enormous influence over the American defense establishment. When he spoke, the military listened. So when Russell reported what he had seen while traveling through the Soviet Union, no one laughed—and hardly anyone outside official circles knew of his remarkable experience until years later.

Just after 7 P.M. on October 4, 1955, while on a train in the Transcaucasia region, the senator happened to gaze out a window to the south. To his considerable astonishment his eyes focused on a large, disc-shaped object slowly ascending as a flame shot from underneath it. The object then raced north across the tracks in front of the train. Russell scurried to alert his two companions, who looked out to see a second disc do what the first had just done. At that moment Soviet trainmen shut the curtains and ordered the American passengers not to look outside.

As soon as they arrived in Prague, Czechoslovakia, the three men went to the United States embassy and sat down with Lieutenant Colonel Thomas S. Ryan, the air attaché. Russell's associate, Lieutenant Colonel E. U. Hathaway, told Ryan that they were about to report something extremely important—"but something that we've been told by your people [the U.S. Air Force] doesn't exist."

Soon rumors about the senator's sighting reached America, but when a reporter for the Los Angeles *Examiner* tried to obtain details, Russell said only, "I have discussed this matter with the affected agencies and they are of the opinion that it is not wise to publicize this matter at this time." The report was not declassified until 1985. Interestingly, one of the "affected agencies" was not Project Blue Book, which never received the report. Apparently, the event was too sensitive for so lowly a project.

"BALL LIGHTNING" IN LEVELLAND

On November 2, 1957, the Soviet Union launched *Sputnik II* into orbit. Within hours, coincidentally or otherwise, a UFO wave erupted in the United States. At first the wave appeared to be concentrated in a small backwater area of west Texas, where a series of remarkable UFO encounters took place.

The sheriff's office in the town of Levelland scoffed that evening when a frightened man called to report that he and a friend, driving on a country highway four miles west of town, had seen a 200-foot-long "rocket" rise up from a field and rush toward their truck. Terrified of an imminent

This is a representation of ball lightning seen during a storm in France in 1845.

collision, the two flew out of the cab and hurled themselves into the ditch. As the UFO passed just above the truck, rocking it with a blast as loud as thunder, the vehicle's engine died and its lights went out, only to resume a few seconds later when the UFO disappeared from view.

An hour later, another caller recounted his experience with an identical UFO that had also interfered with the electric functioning of his car. The scoffing stopped, and sheriff's officers soon found themselves handling comparable stories from frightened observers who had seen a giant, light- and engine-killing UFO at locations west, east, and north of Levelland. At 1:30 A.M. Sheriff Weir Clem and a deputy saw the UFO themselves. A few minutes later Ray Jones, Levelland's fire marshal, experienced motor difficulty when the same or a similar phenomenon was in view.

The official Air Force explanation: "ball lightning." But ball lightning never exceeds more than a few feet in diameter and is usually only inches around. Project Blue Book claimed an electrical storm was in progress during the sightings; there was no storm. By 1957 Project Blue Book's "investigations" were perfunctory at best. Even its chief scientific advisor, astronomer J. Allen Hynek, would later remark on the "absence of evidence that ball lightning can stop cars and put out headlights."

MR. MOORE GOES TO WASHINGTON

Driving near Montville, Ohio, late on the evening of November 6, 1957, Olden Moore was startled to see a glowing disc, 50 feet high and 50 feet in diameter, come down along the roadside. He got out of his car and watched the landed UFO for the next 15 minutes. It was still there when he left to get his wife, but it was gone when they returned. Police and Civilian Defense investigators found both "footprints" and radioactivity at the site.

A few days later Moore disappeared. When he resurfaced, he would not say where he had been. But in private conversations with ufologist C. W. Fitch, Moore claimed that Air Force officers had flown him to Washington, D.C., and hidden him away while they repeatedly interviewed him. Toward the end of his stay, the officers showed him a UFO film, apparently taken from a military plane, and said UFOs seemed to be of interplanetary origin. Moore then signed a document swearing him to secrecy.

DON'T JUDGE A BOOK BY ITS COVER

Judging from the Air Force's press notices, Project Blue Book had the UFO problem well in hand. But in reality, for almost all of its nearly 20-year existence, it was a low-priority operation headed by a lower-ranking officer. A well-funded but highly classified project (even now its name is not known) handled sensitive UFO cases. The staff for Project Blue Book was small and, according to astronomer J. Allen Hynek (Project Blue Book's scientific adviser), less than hardworking. Nonetheless, the Air Force regularly assured reporters, who then uncritically passed the line to newspaper readers, that thorough, scientific investigations had proved the nonexistence of UFOs. In a 1968 letter to the project, Hynek leveled several charges against Project Blue Book: It lacked the trained personnel necessary for the job, had conducted "virtually no dialogue" with the "outside scientific world," and employed statistical methods that were "nothing less than a travesty."

THE DIMENSIONS OF A PHENOMENON

Clyde Tombaugh

OF SAUCERS AND SCIENTISTS

"Have We Visitors from Space?" *Life* magazine asked in an article in its April 7, 1952, issue. It was a question people all over the world were asking in wonder or fear or both. What, short of intruders from other worlds, could explain the presence in the earth's atmosphere of objects that looked like structured craft but performed in ways unimaginably beyond the capacity of earthly rockets and airplanes?

Astronomer Clyde Tombaugh—who had discovered the planet Pluto in 1930—was numbered among those who had seen flying saucers. On the evening of August 20, 1949, he, his wife, and his mother-in-law saw a "geometrical group of faint bluish-green rectangles of light" apparently attached to a larger "structure." He said of the experience, "I have done thousands of hours of night sky watching, but never saw a sight so strange as this."

In 1952, in an informal survey of 44 of his fellow astronomers, J. Allen Hynek of Project Blue Book learned that five had seen UFOs. "A higher percentage than among the public at large," Professor Hynek noted in an internal Air Force memorandum. Fear of ridicule kept most scientists silent about their sightings, however. In a 1976 survey of members of the American Astronomical Society, 62 admitted to having had UFO experiences; only one of the scientists made a public report of his sighting.

One astronomer more than any other would be associated with the UFO phenomenon: Professor

Hynek. As a faculty member at Ohio State University, he was close to Dayton, Ohio, the location of the UFO project's headquarters at Wright-Patterson Air Force Base (AFB). In 1948 the Air Force asked Hynek to look at the UFO reports it was gathering to determine which were merely cases of wrongly identifying astronomical phenomena such as meteors, comets, planets, and stars. To the extent he had given the subject any thought, Hynek was deeply skeptical of flying saucers. Yet four years later, he confessed in a lecture to colleagues that some reports were indeed "puzzling." The "steady flow of reports, often made in concert by reliable observers," merited scientific attention, not ridicule. "Ridicule is not a part of the scientific method," Hynek said, "and the public should not be taught that it is."

Yet Hynek—cautious, even plodding, by nature—did not surrender his skepticism easily. By the late 1950s he was urging his Air Force employers to jettison the term "unidentified flying objects." Reports of UFOs continued to flow in simply because science had failed to educate people to recognize mundane aerial phenomena. Nonetheless, deep down the puzzling cases continued to rankle Hynek, and he watched with growing dismay the clear incompetence of the Project Blue Book "investigation."

By this time Hynek was head of Northwestern University's astronomy department and one of America's best known and most respected astronomers. He had no reason to voice his quiet, heretical concerns that the UFO question had not

J. Allen Hynek in 1962

been satisfactorily answered. But in the early 1960s a graduate student of his, a young Frenchman named Jacques Vallée (who would go on to write a number of UFO books), urged Hynek to give vent to these suspicions, looking at the evidence with an open mind and without fear.

Within a few short years, no one could doubt that even as Hynek remained Project Blue Book's chief scientific adviser, he and the Air Force were now operating on different wavelengths. While the Air Force continued to parrot the same old line—all UFO reports were explainable, and only fools and charlatans thought otherwise—Hynek boldly advocated a new study. With each pronouncement Hynek made less secret his conviction that a new study would show UFOs to be something extraordinary, in all probability the product of a nonhuman intelligence.

Hynek's 1972 book, *The UFO Experience,* eloquently criticized Project Blue Book's and science's neglect of the issue. The book concluded: "When the long-awaited solution to the UFO problem comes, I believe that it will prove to be not merely the next small step in the march of science but a mighty and totally unexpected quantum leap."

NOCTURNAL LIGHTS TO CLOSE ENCOUNTERS

Hynek classified UFO reports into six categories: nocturnal lights, daylight discs, radar/visual cases, and close encounters of the first kind, close encounters of the second kind, and close encounters of the third kind.

Nocturnal lights. Near midnight on the evening of August 30, 1951, during a spate of sightings of boomerang-shaped lights in Lubbock, Texas, college student Carl Hart, Jr., glimpsed a formation of 18 to 20 white lights through his bedroom win-

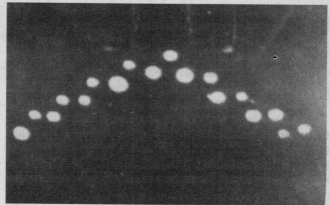

One of the "Lubbock lights" photos taken by Carl Hart, Jr.

dow. Forming a perfect *V* configuration in two rows, they were passing silently over his parents' house from the north. Grabbing a 35mm camera, Hart raced outside, hoping they would return. A minute later the lights reappeared, and though they were visible for less than five seconds, he was able to snap two pictures. When the lights returned once more, Hart got three more pictures.

The local newspaper as well as the Air Force subjected Hart's photographs of the "Lubbock lights"—as they are known in UFO lore—to intensive investigation. No evidence of a hoax emerged then or later, and no conventional explanation could be found.

On January 30, 1993, at 8 P.M., Karen Twilley of Lickskillet, Alabama, took a call from her mother-in-law, who lived across the road from her. She alerted Karen that something strange was outside. Twilley, her husband, and their son stepped outside to see. In one area of the sky, they saw three triangle-shaped lights and in another, a large white sphere. The triangles, though at some considerable distance above them, looked to be larger than airplanes. Each carried red and white blinking lights. "They were in constant motion," Mrs. Twilley related, "but there wasn't any sound." They remained in view for about 45 minutes.

Daylight discs. From mid-morning to mid-afternoon on July 8, 1947, silvery, disc-shaped objects bedeviled Muroc Air Base (later renamed Edwards AFB). Two discs first showed up at 9:30 A.M.; they moved at 300 miles per hour at 8,000 feet altitude on a level flight path against the wind. A third disc, flying in tight circles, then appeared and headed toward the Mojave Desert along with the first two. Some 40 minutes later a test pilot warming up an XP-84 aircraft saw another object, again flying into the wind. At noon, as a pilot was conducting a seat-ejection test at 20,000 feet, observers saw a UFO underneath it. The object was descending rapidly; it then headed north. Witnesses told Air Force investigators that "it presented a distinct oval-shaped outline, with two projections on the upper surface which might have been thick fins or knobs. They crossed each other at intervals, suggesting either rotation or oscillation of a slow type. . . . The color was silver, resembling an aluminum-painted fabric." At 4 P.M. an F-51 pilot encountered a "flat object of a light-reflecting nature" without wings or fins.

These sightings, Hynek wrote, caused the Air Force to "take a deep interest in UFOs."

Reports of these classic "flying saucers" continue up to the present. Early on the evening of September 16, 1990, a young couple on their way home from Sunday dinner at a West Jordan, Utah, restaurant spotted what first looked like a gray wisp of cloud in the western sky. More focused observation revealed it to be a metallic disc with a dome on top. Flying at a 60-degree angle above the horizon, it passed rapidly from the northwest to the southeast and disappeared behind a grove of trees. In view for 30 to 60 seconds, it was the apparent size of two full moons.

As a UFO passed over his farm near McMinnville, Oregon, on May 11, 1950, Paul Trent snapped this picture, which has withstood four decades of investigation and analysis. Nothing has disproved two witnesses' assertion that this was an extraordinary flying object.

Radar/visual. While driving east of Corning, California, near midnight on August 13, 1960, state police officers Charles Carson and Stanley Scott saw a lighted object drop out of the sky. Fearing the imminent crash of an airliner, they screeched to a halt and jumped out of their car. The object continued to fall until it reached about 100 feet altitude, at which point it abruptly reversed direction and ascended 400 feet, then stopped. "At this time," Carson wrote in his official report, "it was clearly visible to both of us. It was surrounded by a glow making the round or oblong object visible. At each end, or each side of the object, there were definite red lights. At times about five white lights were visible between the red lights. As we watched, the object moved again and performed aerial feats that were actually unbelievable."

The two officers radioed the Tehama County Sheriff's Office and asked it to contact the nearest Air Force base (at Red Bluff). Radar there confirmed the object's presence.

The UFO remained in view for more than two hours. During that time two deputy sheriffs and the county jailer saw it from their respective locations. According to Carson:

> On two occasions the object came directly towards the patrol vehicle; each time it approached, the object turned, swept the area with a huge red light. Officer Scott turned the red light on the patrol vehicle towards the object, and it immediately went away from us. We

observed the object use the red beam approximately six or seven times, sweeping the sky and ground areas. The object began moving slowly in an easterly direction and we followed. We proceeded to the Vina Plains Fire Station where it was approached by a similar object from the south. It moved near the first object and both stopped, remaining in that position for some time, occasionally emitting the red beam. Finally, both objects disappeared below the eastern horizon.

Carson noted, "Each time the object neared us, we experienced radio interference."

Between March and May 1990, Russia was inundated with UFO reports. Many were made by soldiers who watched luminous balls, discs, and triangles maneuvering at high altitude and high speed. On March 21, at 9:40 P.M., when one such object appeared at 6,500 feet above the city of Pereslavl-Zalesski, an interceptor piloted by Colonel A. A. Seyonchenko streaked after it. When he caught sight of it, the colonel turned his aircraft's radar on it and tracked the UFO as it flashed two white lights and changed speed and altitude. Seyonchenko flew above the UFO, which did not answer his radio challenges to identify itself; in the darkness he could see only the lights and a vague shape between them.

A ground observer, Captain V. Birin, got a better look at the UFO a few minutes later, as it descended. "The object was shaped like a saucer with two bright lights on the edges," he said. "The diameter was about 100 to 200 meters. There was dim illumination between the lights looking like portholes."

From 8:00 until just after 10:00 that evening, numerous witnesses, both civilian and military, observed this and other UFOs. The result was, in the words of investigator V. D. Musinsky, "the most detailed, precise, and documented report in the history of ufology. The commanding officers of several antiaircraft defense units around Moscow have gathered more than 100 visual reports from their subordinates to complete the picture."

Close encounters of the first kind. Hynek defines one of these as "a close-at-hand experience without tangible physical effects."

A couple was driving north on Highway 45 north of Bristol, Wisconsin, at 11 P.M. on October 14, 1986. They saw flashing red and white lights, which they took to mean that a car accident had occurred on the road just ahead of them. Approaching cautiously, they were stunned to find the real cause: an enormous, triangular-shaped object hovering just above the concrete. The lights ran along the object's outer edge. "It was the size of a two-story house and spanned the width of the road," the husband told Don Schmitt of the Center for UFO Studies (CUFOS), the organization Hynek founded in 1973.

The couple stepped out of the car and gazed up at the structure, no more than 20 feet above them. On its bottom they could make out a grid structure. Two minutes later the object drifted off to the southeast and was lost to view. They told Schmitt, "It was so low that if we would have stood on the roof of the car, we could almost have touched it."

Passing through the blueberry barrens of Washington County, Maine, in the late afternoon darkness of November 2, 1990, Warren and Velma Orcutt noticed lights shining through the trees. At first they thought the lights emanated from a nearby town, but when they turned a bend in the road, they observed the true source: a saucerlike object 60 feet in diameter. Just ahead of them and slightly to their left, it was rising slowly, its bottom section partially enshrouded in mist. Through the fog the Orcutts could see three brilliant, swiveling lights, red, green, and white, moving in a counterclockwise direction and illuminating the ground and trees below. These lights were so bright that they hurt the witnesses' eyes. The UFO rose, then stopped and hovered before rising again. Its motion led the Orcutts to suspect it was "trying to avoid striking the limbs of the trees."

Orcutt drove his truck a few hundred feet down the road to a higher elevation. He and his wife turned off the truck's lights and engine and watched as the UFO continued to hover. It made no sound. Now the two could see that the object had rectangular-shaped "windows" from which a dull, orange glow emanated.

A few minutes later the Orcutts drove to a nearby house and alerted its occupants to the strange object's presence. Together with members of this family, they watched the UFO's slow ascent until finally it shot away at "terrific speed" along a horizontal course to the northeast.

Close encounters of the second kind. In this kind of UFO encounter "a measurable physical effect on either animate or inanimate matter is manifested."

Late on the afternoon of January 8, 1981, at Trans-en-Provence, France, a whistling sound disturbed Renato Nicolai as he worked in his garden. When he saw a lead-colored "ship" moving toward him from two pine trees at the edge of his property, he fled to a small cabin on a nearby hill. From there Nicolai saw the object, shaped like "two saucers upside down, one against the other," descend to the ground. Shortly thereafter it rose up and shot off toward the northeast. On its bottom Nicolai observed "two kinds of round pieces which could have been landing gear or feet."

Not long afterward the gendarmerie appeared on the scene and wrote in their official report: "We observed the presence of two concentric circles, one 2.2 meters in diameter and the other 2.4 meters in diameter. The two circles form a sort of corona ten centimeters thick on this corona, one within the other. There are two parts clearly visible, and they also show black striations." Groupe d'Étude des Phénomènes Aerospatiaux Non-Identifiés (GEPAN), France's official UFO-investigative agency, took soil and plant samples to the nation's leading botanical laboratory.

After a two-year study GEPAN determined that a "very significant event . . . happened on this spot." GEPAN head Jean-Jacques Velasco wrote, "The effects on plants in the area can be compared to that produced on the leaves of other plant species after exposing the seeds to gamma radiation." In its 66-page technical monograph on the case, GEPAN cautiously acknowledged that the incident amounted to proof that a UFO had landed: "For the first time we have found a combination of factors which conduce us to accept that something similar to what the eyewitness has described actually did take place."

When a dull, "throbbing" sound woke up a Jupille, Belgium, man at 2:15 A.M. on December 12, 1989, he grabbed a flashlight and went to investigate. Outside he found its source: a large, metallic, oval-shaped craft hovering between trees along a nearby road.

Along the UFO's surface, clearly visible in the bright moonlight, the witness could see a logo that looked like several ellipses crossing themselves. Small lights changing from blue to red and back again flashed along the UFO's circumference. In the rear was an oar- or paddle-shaped device; in front was a cockpit window. The whole effect, the witness thought, was like something "out of one of the novels of Jules Verne."

After several minutes the UFO ascended a small distance, then headed slowly in the direction of a neighbor's meadow. As it moved, it directed searchlight beams toward the ground. It disappeared behind the neighbor's house, at one point shooting a shaft of light up into the sky. By now badly frightened, fearing he was seeing something he was not meant to see, the witness decided to go back to bed.

On awakening a few hours later, he notified the gendarmerie, who soon arrived in the company of army representatives and local police officers. A gigantic circular impression could be seen in the meadow. In the middle of the impression the grass had been cut, but no grass cuttings could be found. The grass within the trace had turned yellow. The investigators seemed curiously uninterested in what the witness had to say about his sighting. It was "as if they knew what it was all about," he would recall. They cordoned off the area as they collected samples and took photographs.

Officials have never released their report on the case. Civilian researchers learned that two other witnesses in the area heard the throbbing sound on the night in question, and a local journalist had seen an unusual light but had not gone outside to determine where it came from.

This event took place during a spectacular 1989–90 sighting wave over Belgium and France.

Close encounters of the third kind. In an occurrence of this type, "the presence of animated creatures is reported" inside or in the vicinity of UFOs.

As William Squyres, a musician at a Pittsburg, Kansas, radio station, drove to work at 5:50 A.M. on August 25, 1952, he encountered a large disc hovering ten feet above the ground about 250 yards away. He quickly brought his car to a stop, jumped out, and began walking toward the UFO. It looked, he later told Project Blue Book investigators, like two bowls placed end on end, 75 feet long and 40 feet wide, with a 15-foot-high midsection. Along the side was a row of windows.

Through these windows Squyres detected movement of some sort, but he could not detect its cause. In one window he could see the head and shoulders of a humanlike figure who seemed to be leaning forward and watching him. The UFO departed before Squyres could get any closer to it. As it ascended, according to the Project Blue Book report, "it made a sound like a large covey of quail starting to fly at the same time."

This incident is among the few UFO reports Project Blue Book acknowledged it could not explain.

Late in February 1991 former police officer Luis Torres was camping on El Yonque mountain in the Carribbean National Forest of Puerto Rico with his wife, Marguerita, and two policemen and their wives. Still awake at 3:15 A.M., they saw a bizarre sight—"two little persons coming down the highway in front of the ranger station," Torres told ufologist Jorge Martin.

His wife recalled, "They were like two little boys, two little men three or four feet tall. Large heads and black eyes. It was as if they had large black circles where their eyes were. To me they weren't human; they were some bizarre creatures. They had their own language, a rapid buzzing sound. They never looked at us directly."

Luis Torres estimated their height to be four feet. "In the moonlight, and there wasn't much moonlight that night," he said, "the clothing looked monochromatic, a smooth, even greenish gray color. It went over their heads, covered the top of their skulls. The only skin that was exposed was the face and hands, nothing more. Their arms were long, to the knees, and their hands were long, too, as near as we could tell. . . . The skin appeared to be a grayish green.

"They went down the highway talking to each other in that funny language, and when they were some hundred feet beyond us they did an about-face and started back toward us."

At that point Torres drew his pistol. "When I took out the weapon, they realized it. They did not look directly at us at any time, but they started to walk faster, and when they got just beyond us they crossed the highway to the left and jumped into the thick underbrush there, where the ground is steep. My brother-in-law and I left running to follow them, but they lost us in the underbrush when we were about 60 feet from them."

This is only one of a number of reports of similar beings that have been recorded during an ongoing UFO wave in Puerto Rico. Torres says, "I don't drink or smoke, nor do the others. We are religious people, and we're sure of what we saw. They were little things that looked like people, but they were not human beings."

ALIEN AUTOPSIES

There is no shortage of anecdotes, intriguing but unverifiable, of alien bodies glimpsed in top secret U.S. government laboratories and warehouses. Veteran ufologist Leonard H. Stringfield of Cincinnati has collected numerous such reports from the late 1940s to the present. Stringfield has published these in a series of monographs on what he calls "retrievals of the third kind"—recoveries of crashed UFOs and their (usually) dead occupants.

In 1978 Stringfield learned of a physician who claimed to have performed autopsies on an alien being in the early 1950s. Stringfield established that the man was a doctor on the staff of a major hospital, but the informant insisted that his name be kept secret. In a statement provided for Stringfield in July 1979, he wrote:

> The specimen observed was four feet three inches tall in length. I can't remember the weight. It has been so long and my files do not contain the weight. I recall the length well, because we had a disagreement and everyone took their turn at measuring. The head was pear-shaped in appearance and oversized by human standards for the body. The eyes were Mongoloid in appearance. The ends of the eyes farthest from the nasal cavity slanted upward at about a 10-degree angle. The eyes were recessed into the head. There seemed to be no visible eyelids, only what seemed like a fold. The nose consisted of a small foldlike protrusion above the nasal orifices. The mouth seemed to be a wrinklelike fold. There were no human lips as

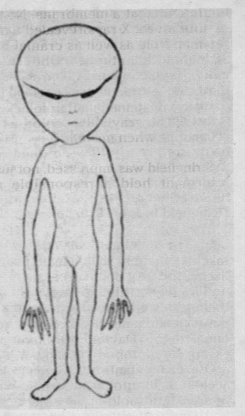

A drawing based on the doctor's reports.

such—just a slit that opened into an oral cavity about two inches deep. A membrane along the rear of the cavity separated it from what would be the digestive tract. The tongue seemed to be atrophied into almost a membrane. No teeth were observed. X rays revealed a maxilla and mandible as well as cranial bone structure. The outer "earlobes" didn't exist. The auditory orifices present were similar to our middle and inner ear canals. The head contained no hair follicles. The skin seemed grayish in color and seemed mobile when moved.

Stringfield was impressed, not just because his informant held a responsible position, but because Stringfield had other stories that described similar alien bodies.

Ufologist Isabel Davis interviewed a medical doctor who claimed that in the late 1950s she was taken to a highly secure facility to study human-like portions of bodies. She saw they came from no known earthly species. Told nothing about them, she was instructed to write her conclusions down and then never discuss the matter with anyone. She told Davis the admonition was unnecessary, since nobody would believe her.

Other informants tell of seeing, intentionally or accidentally, autopsy reports based on examinations of humanoid bodies. Such reports are fantastic and unprovable. Yet at least some come from individuals who seem sincere and sane. Moreover, their descriptions are sometimes strikingly consistent. Until it can be proven, however, that alien bodies exist in some military hiding place, such tales will remain just tales.

CASE STUDIES

BURNED BY A UFO

Stephen Michalak was searching for minerals along Falcon Lake, 80 miles east of Winnipeg, Manitoba, on May 20, 1967, when he heard the cackling of geese. Looking up into the early afternoon sky, he saw two glowing oval-shaped objects on a steep, swift descent. One abruptly stopped its downward flight while the other object continued, landing on a flat rock outcropping 160 feet away.

Michalak carefully approached the strange craft, which was 40 feet wide and 15 feet high. It looked like a bowl with a dome on top and emitted a humming sound and a sulphur stench. On the bottom half, just below the rim of the bowl, was a doorlike opening from which Michalak heard muffled voices. "They sounded like humans," he reported. "I was able to make out two distinct voices, one with a higher pitch than the other."

Quite naturally, Michalak thought he was dealing with a terrestrial craft. He addressed the speakers in several languages, asking if he could help. However, he got no answer. He poked his head through the opening into the interior, seeing only a "maze of lights." At that moment three panel doors slid across and sealed the opening. As Michalak stepped back, he touched the vehicle's exterior: It was so hot that it burned his gloves.

Suddenly the object rose, expelling hot air through a gridlike vent and causing Michalak's shirt to erupt into flames. An attack of nausea overtook him.

When doctors examined Michalak in a Winnipeg hospital a few hours later, they noted a dramatic burn pattern all across his chest—exactly like the grid Michalak had described on the UFO's underside. Michalak's health problems continued and brought him to Minnesota's Mayo Clinic the next year. Investigations by official and civilian bodies uncovered no evidence of a hoax. As late as 1975, a member of the Canadian Parliament complained that the government had not released all its findings.

A UFO FOR THE PRESIDENT-TO-BE

One night in 1974, from a Cessna Citation aircraft, one of America's most famous citizens saw a UFO.

The plane held four persons: pilot Bill Paynter, two security guards, and the governor of California, Ronald Reagan. As the airplane approached Bakersfield, California, the passengers called Paynter's attention to a strange object to their rear. "It appeared to be several hundred yards away," Paynter recalled. "It was a fairly steady light until it began to accelerate. Then it appeared to elongate. Then the light took off. It went up at a 45-degree angle—at a high rate of speed. Everyone on the plane was surprised. . . . The UFO went from a normal cruise speed to a fantastic speed instantly. If you give an airplane

*People of all kinds have seen UFOs, including Ronald
Reagan when he was governor of California.*

power, it will accelerate—but not like a hot rod,
and that's what this was like."

A week later Reagan recounted the sighting to
Norman C. Miller, who was then Washington
bureau chief for *The Wall Street Journal.* Reagan
told Miller, "We followed it for several minutes. It
was a bright white light. We followed it to
Bakersfield, and all of a sudden to our utter
amazement it went straight up into the heavens."
When Miller expressed some doubt, a "look of
horror came over [Reagan]. It suddenly dawned
on him . . . that he was talking to a reporter."
Immediately afterward, according to Miller,
Reagan "clammed up."

Reagan has not discussed the incident publicly
since.

ANGEL HAIR

It may have been the strangest sight ever to grace the sky over Oloron, France. In the early afternoon of October 17, 1952, according to high school superintendent Jean-Yves Prigent—who was only one of the many witnesses—a "cottony cloud of strange shape" appeared overhead. "Above it, a narrow cylinder, apparently inclined at a 45-degree angle, was slowly moving in a straight line toward the southwest. . . . A sort of plume of white smoke was escaping from its upper end."

In front of this "cylinder" were 30 smaller objects that when viewed through opera glasses proved to be red spheres, each surrounded by a yellow ring. "These 'saucers' moved in pairs," Prigent said, "following a broken path characterized in general by rapid and short zigzags. When two saucers drew away from one another, a whitish streak, like an electric arc, was produced between them."

But this was only the beginning of the strangeness. A white, hairlike substance rained down from all of the objects, wrapping itself around telephone wires, tree branches, and the roofs of houses. When observers picked up the material and rolled it into a ball, it turned into a gelatinlike substance and vanished. One man, who had observed the episode from a bridge, claimed the material fell on him, and he was able to extract himself from it only by cutting his way clear—at which point the material collected itself and ascended.

This artist's representation shows the phenomenon seen by many residents of Oloron, France: the cylinder surrounded by the red spheres, which rain down the white, cottony substance known as "angel hair."

A nearly identical series of events occurred in Gaillac, France, ten days later.

Such "angel hair" is reported from time to time, although airborne cobwebs are sometimes mistaken for angel hair. Laboratory analysis of authentic material is impossible because the material always vanishes before it can be analyzed. In the summer of 1957, when Craig Phillips (director of the National Aquarium from 1976 to 1981) witnessed a fall off the Florida coast, he collected samples and placed them in sealed jars. But by the time he got to his laboratory, they were gone.

ALASKA MOTHERSHIP

When he first saw them, Japanese Airlines officer Kenju Terauchi, who was piloting a Boeing 747 cargo plane, thought they were lights from a military aircraft. He soon learned otherwise. During the next half hour he and his crew realized that things of a decidedly unearthly character had joined them in the skies over Alaska. It was November 17, 1986, at 5:10 in the afternoon.

The pilot, first officer, and flight engineer saw two lighted structures, "about the same size as the body of a DC-8 jet," in Terauchi's words, moving about 1,000 feet in front of the cargo craft. Terauchi's radio communications to Anchorage flight control were strangely garbled, but enough got through that Anchorage urgently contacted a nearby Air Force radar station to see what it was picking up. At various times during the event the UFOs were tracked by the 747 on-board radar and by the Air Force ground radar.

As the sky darkened, the UFOs paced the 747 and were finally lost in the distant horizon. Then a pale white light appeared behind the aircraft. Silhouetted against lights on the ground, it looked like an immense, Saturn-shaped object—the size, Terauchi estimated, of "two aircraft carriers." He thought it was a "mothership" that had carried the two "smaller" objects, themselves of no inconsequential size. The Anchorage radar was recording the object's presence. For the first time the crew felt fear. By now the aircraft was running low on fuel, and the captain requested permission to land. The UFO vanished suddenly at 5:39 P.M.

LANDING AND TRACES AT VALENSOLE

Near the French village of Valensole, farmer Maurice Masse was smoking a cigarette just before starting work at 5:45 A.M. on July 1, 1965. An object came out of the sky and landed in a field of lavender 200 feet away. Annoyed, and assuming that a helicopter had made an unauthorized landing, he walked toward it. However, he soon saw it was no helicopter but an oval-shaped structure resting on four legs. In front of it stood two figures, not quite four feet tall, dressed in tight, gray-green clothes. Their heads were oversize and had sharp chins, their eyes were large and slanted, and they were making a "grumbling" noise.

One of the beings pointed a pencillike device at Masse, paralyzing him in his tracks. The figures entered the UFO and flew away. Masse needed 20 minutes to recover his mobility. In its wake the object had left a deep hole and a moist area that soon hardened like concrete. Plants in the vicinity decayed, and analysis found a higher amount of calcium at the landing site than elsewhere.

The Valensole case is considered one of the classic UFO incidents. Investigations by official and civilian agencies confirmed Maurice Masse's sincerity and good character. Laboratory study of the affected soil and plants confirmed the occurrence of an unusual event. Subsequently Masse confided that in the course of the encounter he experienced some sort of communication with the entities.

HOT ENCOUNTERS

During the great sighting outbreak of early November 1957, a number of close encounters had a disturbing consequence: burns and related injuries to witnesses. One of the most dramatic occurrences took place at an army base at Itaipu along Brazil's Atlantic coast. At 2 A.M. on November 4, two guards saw a luminous orange disc coming in over the ocean at a low altitude and an alarming rate of speed. As it passed above the soldiers, the disc came to an instant stop. The

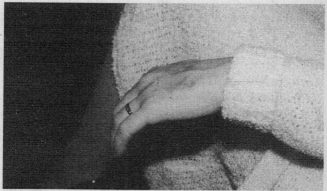

In 1981 a woman in England was burned on the hand by a beam coming from a large, disc-shaped object.

two witnesses suddenly felt a wave of heat and a horrifying sensation as if they had burst into flame. Their screams brought other soldiers stumbling out of their barracks just in time to see the UFO streak away. At that moment the fort's entire electrical system failed. Amid great secrecy the two men were rushed to a military hospital and treated during the next several weeks for first- and second-degree burns to ten percent of their bodies.

But there were other burn cases as well. In the afternoon of the same day as the Itaipu incident, the engines of several cars along a rural highway near Orogrande, New Mexico, ceased to function as an egg-shaped object maneuvered close by. A witness who stood particularly close to it contracted a "sunburn." In the early morning hours of November 6, outside Merom, Indiana, a hovering UFO bathed René Gilham's farm in light and also seriously burned his face. He ended up spending two days in the hospital. At around 1:30 A.M. on November 10, a Madison, Ohio, woman saw an acorn-shaped UFO hovering just behind her garage. She watched it for half an hour. In the days afterward she developed a body rash and vision problems that her doctor believed suggested radiation poisoning. Subsequent medical tests uncovered no apparent cause for her injuries.

THE LITTLE MEN OF NORTH HUDSON PARK

At around 2:45 A.M. on January 12, 1975, George O'Barski was driving home through North Hudson Park, New Jersey, just across the Hudson River from Manhattan, when static filled his radio. Leaning forward to fiddle with the dial, he noticed a light to his left. A quick glance, followed by an astonished stare, revealed its source: a dark, round object with vertical, brilliantly lit windows. It was heading in the same direction as the car and emitted a humming sound.

O'Barski slowed down so he could get a better view. The UFO entered a playing field and hovered a few feet off the ground. A panel opened between two windows, and a ladder emerged. Seconds later about ten identically clad little figures—they wore white, one-piece outfits with hoods or helmets that obscured their facial features—came down the ladder. Each figure dug a hole in the soil with a spoonlike device and dumped the contents into a bag each carried. The figures then rushed back into the ship, which took off toward the north. The entire incident had lasted less than four minutes.

Months later O'Barski confided the story to a longtime acquaintance, Budd Hopkins, who was interested in UFOs. Hopkins and two fellow investigators subsequently found independent witnesses who verified the presence of a brightly lit UFO in the park at the time of O'Barski's sighting, although only O'Barski was close enough to see the little figures. One witness, a doorman at an apartment complex bordering the park, said that as he watched the object, he heard a high-pitched vibration, and the lobby window broke just as the UFO departed.

DAMAGED CAR IN MINNESOTA

Studying the brilliant light in the stand of trees two and a half miles south of him, Marshall County Deputy Sheriff Val Johnson wondered if drug smugglers had flown over the Canadian border into the flat, isolated terrain of far northwestern Minnesota. The light was close to the ground, suggesting that the plane had either landed or crashed. Or maybe there was some simpler explanation. Johnson headed down the county highway to investigate. It was 1:40 A.M. on August 27, 1979.

The next thing Johnson knew, the light was shooting directly toward him, moving so fast that its passage seemed instantaneous. The last thing he heard was the sound of breaking glass.

At 2:19 A.M. a weak voice crackled over the radio in the sheriff dispatcher's office at Warren, Minnesota. It was Johnson, who had just regained consciousness. His car had skidded sideways and now was stretched at an angle across the northbound lane, its front tilting toward the ditch. When asked what happened, Johnson could only reply, "I don't know. Something just hit my car."

Officers who arrived on the scene found the car had sustained strange damage, including a seriously cracked windshield, a bent antenna, smashed lights, and other damage. Both the car clock and Johnson's wristwatch were running 14 minutes slow, though both had been keeping correct time until the UFO incident. Johnson's eyes hurt badly as if, an examining physician declared, from "welding burns."

An extensive investigation was conducted by Allan Hendry of the Center for UFO Studies along with experts from Ford and Honeywell. Their conclusion: The windshield damage was caused by stones apparently carried in the wake of the unknown object. The Honeywell expert thought the bent antenna probably resulted from a "high-velocity air blast superimposed on the air movement over the fast-moving car."

THE CASE OF THE MISSING WATER

The sounds of frightened cattle woke a rancher from a sound sleep in the early morning of September 30, 1980, near Rosedale, Victoria, Australia. When he went outside, he was astonished to see a domed disc with orange and blue lights gliding about ten feet above the ground. It rose slightly in the air, hovered briefly above an open 10,000-gallon water tank, and then landed 50 feet away. The rancher jumped on a motorbike and sped toward the object, which was making a "whistling" sound. Suddenly an "awful scream" sounded as a black tube extended from the UFO's base. With an ear-splitting bang, the strange craft rose into the air. A blast of hot air almost knocked the witness down.

The sounds ceased as the object slowly moved to a position about 30 feet away and eight feet above the ground. Hovering briefly, it dropped debris—stones, weeds, cow dung—from underneath it, then flew away, disappearing in the east.

A ring of black, flattened grass 30 feet in diameter marked the place where the disc had landed. When he examined it in the daylight, the witness discovered that all the yellow flowers within the circle had been removed. Only green grass remained. But even more bizarre, the water tank was empty, with no evidence of spillage. Only the muddy residue at the bottom of the tank was left—and it had been pulled into a two-foot-high cone shape. The witness was sick with headaches and nausea for more than a week afterward.

A similar ring was found the following December at Bundalaguah, not far from Rosedale. The water in a nearby reservoir was also mysteriously missing.

OVER THE RAINBOW

In *UFO Reality* (1983) British ufologist Jenny Randles noted that some UFO witnesses experience a "sensation of being isolated, or transported from the real world into a different environmental framework. . . . I call this the 'Oz Factor,' after the fairy-tale land of Oz."

In one instance on a late-summer evening in 1978, a Manchester, England, couple watched a UFO hover above a usually well-traveled street almost inexplicably devoid of its customary brisk vehicular and pedestrian traffic. On the afternoon of April 15, 1989, a father and son watched a metallic, gold-colored, dumbbell-shaped object, accompanied by four smaller discs, maneuvering low in the sky near their home in Novato, California. As puzzling to the witnesses as the UFOs themselves was the absence of other humans at a time of day when people would ordinarily be out.

In Randles' view such reports suggest that in some way the "consciousness of the witness [is] the focal point of the UFO encounter."

"SPACE BROTHERS," MUFON, AND CUFOS

Most scientists have been apathetic to UFO reports, and official government agencies have treated them with indifference or wrapped them in secrecy. As a result we owe most of our UFO knowledge to the efforts of private organizations of varying credibility. At one extreme are quasi-religious groups advocating contact with godlike "Space Brothers." At the other are sober, scientifically grounded efforts that carefully gather and document evidence while keeping speculation to a minimum.

The three most influential groups in the United States have been the Aerial Phenomena Research Organization (now defunct), the Mutual UFO Network (103 Oldtowne Road, Seguin, Texas 78155), and the Center for UFO Studies (2457 West Peterson Avenue, Chicago, Illinois 60659). MUFON and CUFOS both publish magazines *(MUFON UFO Journal* and *International UFO Reporter)* that cover a wide range of UFO matters.

THE ABDUCTION ENIGMA

The most famous case of UFO abduction is Betty and Barney Hill's 1961 encounter in New Hampshire.

On the evening of September 19, 1961, while driving home to Portsmouth through rural New Hampshire, Barney and Betty Hill sighted a pancake-shaped UFO with a double row of windows. At one point they stopped their car, and Barney got out for a better look. As the UFO tilted in his direction, he saw six uniformed beings inside. Suddenly frightened, the Hills sped away, but soon a series of beeps sounded, their vehicle started to vibrate, and they felt drowsy. The next thing they knew, it was two hours later than they expected; somehow the Hills had lost two hours.

A series of disturbing dreams and other problems led the Hills to seek psychiatric help. Between January and June 1964, under hypnosis, they recounted the landing of the UFO, the emergence of its occupants, their abduction into the craft, and their separately experienced medical examinations. In 1965 a Boston newspaper reported the story, which in 1967 became the subject of a best-selling book, *The Interrupted Journey.* On October 20, 1975, NBC television broadcast a docudrama, *The UFO Incident,* about the experience.

Nearly everyone has heard of the UFO abduction of the Hills. At the time it shocked even hardcore ufologists. Nothing quite like it had been recorded. Ufologists did know of a bizarre December 1954 incident from Venezuela: Four hairy UFO beings allegedly tried to drag a hunter into their craft, only to be discouraged when his companion struck one of them on the head with the butt of his gun. Ufologists traditionally viewed with suspicion claims of on-board encounters with UFO crews. Those kinds of stories were associated with "contactees," who were regarded, with good reason, as charlatans who peddled long-winded tales of meetings with godlike "Space Brothers." The Hills, however, had sterling personal reputations, and they returned from their experience with no messages of cosmic uplift.

Yet as the 1960s went on, it was becoming unmistakably clear that a new element had entered the UFO picture. Even the Air Force–sponsored Condon committee (see Chapter 2, "UFOs: The Official Story") came upon a similar case in the course of its investigations. It involved an Ashland, Nebraska, police officer named Herbert Schirmer. While patrolling the outskirts of town in the early morning hours of December 3, 1967, he encountered a football-shaped object above the highway not 40 feet from him.

This sighting was all that Schirmer consciously remembered. He remained puzzled by a 20-minute lapse he could not account for, and he suffered from headaches and anxieties associated with the encounter. Committee investigators had him placed under hypnosis, during which he told of being taken aboard the craft and briefly communicating with the crew commander, described as slender, five feet tall, thin-faced, gray-skinned, and generally humanlike.

The 1970s saw the proliferation of such reports. An October 11, 1973, encounter by two Pascagoula, Mississippi, fishermen, who claimed to have been taken into a craft by robotlike entities,

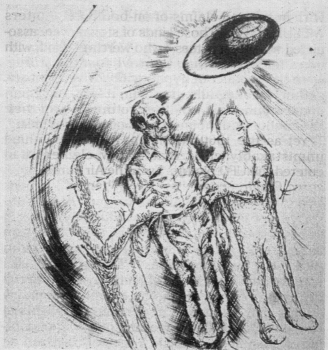

This drawing illustrates the abduction of one of the two fishermen, Charles Hickson.

attracted national attention. So did two abduction reports in 1975—Sandy Larson's in North Dakota (see page 105) and Travis Walton's in Arizona (see page 107). Meanwhile ufologists were studying other incidents, and in 1977 the first book devoted exclusively to the subject—*Abducted!*, by Coral and Jim Lorenzen—appeared.

HYPNOSIS AND CONFABULATION

The stories seemed so fantastic that even many ufologists doubted them. In 1977 three California investigators, seeking to prove that the stories were the product of hypnotically driven fantasy, gathered college-student volunteers, put them under hypnosis, and directed them to imagine UFO abductions. The result was disappointing. The researchers had expected to find "real" and imaginary abductions indistinguishable, but in fact they shared few features in common.

Still, caution about testimony rendered under hypnosis was by no means unwarranted. Contrary to popular misunderstanding, hypnosis is no royal road to the truth. Subjects of hypnotism are in a highly suggestible state and may seek to please the hypnotist. Thus if the hypnotist asks leading questions, the subject may provide the desired answers. Moreover, purely imaginary events can seem real under hypnosis; this is called confabulation. An example is the recounting of "past lives" while in a hypnotic state.

To test the confabulation hypothesis, folklorist Thomas E. Bullard collected all abduction accounts available by the mid-1980s. He found that as many as one-third of the informants had full conscious recall of their experiences and had never resorted to hypnosis to elicit the details. These nonhypnosis reports proved identical in all significant particulars to those told under hypnosis. Bullard also learned that the identity of the individual hypnotist made no difference. The stories remained consistent down to details that even those most familiar with the phenomenon had failed to notice. In short, Bullard concluded, whatever its ultimate cause, the abduction phenomenon was not the product of hypnosis. "The skeptical argument needs rebuilding from the ground up," he wrote.

FANTASTIC ENCOUNTERS

More than any other investigator, Budd Hopkins has shaped current ideas about the nature of the abduction phenomenon. Hopkins, an artist who lives in New York City, has written two widely read books, *Missing Time* (1981) and *Intruders* (1987), and has lectured all over the world. His findings have been widely replicated, though controversy continues about what these findings mean.

According to Hopkins, most abductees have a lifelong association with the abductors. Such experiences begin in young childhood. The majority of abductees encounter small, minimally communicative humanoids with big heads, large and slanting eyes, and thin, gray bodies. Abductees are subjected to intrusive physical examinations, and sometimes tiny "implants" are inserted into their bodies. The abductors have a particular interest in human emotions, possibly because they themselves are largely cold and unfeeling.

Another focus of fascination is human reproduction. In the most fantastic, disturbing, or unbelievable accounts (take your pick), female abductees report pregnancies, some of seemingly inexplicable origin. They also describe the abrupt and equally inexplicable termination of these pregnancies. Under hypnosis the women link these occurrences to abductions initiated to impregnate them, then other abductions to remove the fetus. Several years later, in subsequent abduction experiences, the women are shown children who they are led to believe are their own, human-alien hybrids. Male abductees recount episodes of sexual intercourse with more or less human-looking females or of having sperm samples removed by a device being placed over their genitals.

David M. Jacobs, an associate professor of history at Temple University in Philadelphia, has expanded on Hopkins' work in his own investigations, through hypnosis, with abductees. According to Jacobs, the abductors are subject to a command structure, with a larger but otherwise similar being in charge. After the smaller beings conduct a physical examination of the abductee, the taller being enters, stares closely into the abductee's eyes with his own large, dark, almond-shaped eyes, and extracts information from, or creates images in, the abductee's head.

Though such encounters often terrify and traumatize abductees, their captors usually treat their suffering with at best polite indifference. Moreover, the beings, Jacobs writes in *Secret Life* (1992), "express absolutely no interest in anything about the abductee's daily life apart from physiology. They express no interest in her personal, social, or family relationships, except as they bear upon the breeding program. They express no interest in politics, culture, economics, or the rich and extraordinarily complex tapestry that makes up human relationships and societies. They do, however, express interest in birth control, smoking, and health problems that might directly relate to childbearing for women."

The major scientific proponent of the abduction phenomenon, psychiatrist and Pulitzer Prize–winning author John E. Mack of the Harvard Medical School, accepts the reality of such experiences, though in his view the intentions of the alien intelligences are benevolent. The aliens may be here to guide us into a greater understanding of ourselves and, beyond that, the larger cosmic reality implied in UFO visitation.

SEARCHING FOR THE TRUTH

Could such fantastic claims be true?

The further one is from the question, the more certain the answer: that these stories are all nonsense—the ravings of mentally disturbed individuals or attention-seeking tellers of tall tales. But these glib dismissals, already the subject of a growing body of debunking literature, are held by no one who has actually interacted with individual abductees and examined the testimony up close.

At the same time, the evidence that actual alien kidnappings are occurring is at best circumstantial. Claims as extraordinary as these (which, if true, would shake not only science but all of human life to its foundations) need far better than that. Current evidence is enough to establish the existence of a *phenomenon* as opposed to a random assortment of delusions grouped loosely under the heading "UFO abductions." This phenomenon displays certain features that are consistent over time and space. Bullard, for example, determined that the stories break down into a maximum of eight chronologically consistent elements. For this and other reasons he concluded, "At least *something* goes on."

This "something" includes episodes of genuinely missing time that mere absent-mindedness cannot account for. Even multiple-witness cases include missing time. Family members or friends of the abductees verify their inexplicably late arrivals and attendant confusion. In cases of more than one person being abducted at a time, their descriptions of entities, procedures, and UFO interiors are consistent. In other cases only one person of a group may be taken; the others are "switched off"—left in a state of oblivious paralysis. Later they recall only seeing a strange object initially and then subsequently missing time.

Abductees may also have unusual body scars. Frequently they have no conscious memory of how they came to have these scars; their testimony under hypnosis links them to surgical procedures used by their alien captors.

Dr. Mack writes of attempts to explain away the abduction phenomenon, "Psychosocial hypotheses . . . are difficult to take seriously. For unless we are willing to extend our notions of the powers of the psyche to include the creation of cuts, scars, hemorrhages, and bruises, the simultaneous production of highly elaborate and traumatic experiences similar to one another in minute detail among individuals who have not communicated with one another, and all of the physical phenomena associated with the UFOs themselves, such explanations appear quite inadequate."

The scars associated with abduction experiences are not just physical ones. A growing number of mental health professionals have conducted their own investigations of the abduction phenomenon. They insist it is unlike any psychological aberration ever recorded. Its effects, however, are measurable.

In the early 1980s psychological testing of a small group of abductees in New York indicated that they suffered from post-traumatic stress disorder (PTSD). This disorder was first documented and named by John P. Wilson from his study of Vietnam War veterans. Dr. Elizabeth Slater, a psychologist with a private practice in New York City, remarked that the findings are "not inconsistent with the possibility that reported UFO abductions have, in fact, occurred." Wilson and other psychologists have written of the real trauma exhibited by abductees.

Available evidence does not allow any firm conclusions to this intriguing mystery. In fact, the lack of such evidence in some cases is spectacular. Though finding abductees who describe "phantom pregnancies" is not hard, not a single medically documented case has ever been produced. If these events are occurring by the hundreds, thousands, or even (as some abduction proponents maintain) tens of thousands, why has nothing ever been published in the pediatric literature? One would think physicians would find such a pregnancy a baffling enigma. Dr. Richard Neale, a California physician who is interested in the UFO phenomenon, has actively searched for such cases, so far without success.

Claims of abductees becoming pregnant existed before Budd Hopkins came along to write about them, but only one UFO writer, John A. Keel, paid any attention to them. In his 1975 writings, however, he mentioned them only briefly and dismissed them as "hysterical pregnancies." Whatever they are, at this time they exist only as curious anecdotes. To take seriously the claims of abductees and their proponents, actual medical case histories have to be produced.

Controversial UFO theorist John Keel

Another problem is that there are *too many* abduction stories. In 1992 some abduction researchers interpreted a commissioned survey conducted by the Roper Organization to indicate that some five million persons *in the United States alone* are "probably" abductees. This figure is in no sense credible for any number of reasons—not the least of them flaws in the design of the survey and in the analysis drawn from its data. Even if the survey had been methodologically perfect, the presence of five million supposed abductees would be better explained in psychological terms than in extraterrestrial ones.

As critic Dennis Stacy caustically observed, "If we are to assume that one in every 50 people on a planet with a population of several billion has actually been abducted at one time or another, we are now looking at a potential body count of some several hundred millions. The logistics of an ongoing extraterrestrial invasion on that kind of scale simply won't compute. If true, in fact, the earth's skies would literally be darkened with abducting UFOs day and night; they would be stacked up over the major metropolitan areas in the same way that our own 747s now crowd the air lanes over New York, Chicago, Paris, London, Frankfurt, Singapore, Tokyo, and numerous everyday ports of call."

In fact, the abduction experience—whatever its cause—is probably rarer than even more conservative figures suggest. "Psychosocial Characteristics of Abductees," a psychological inventory of abductees conducted by three social scientists, Mark Rodeghier, Jeff Goodpaster, and Sandra Blatterbauer, appeared in *Journal of UFO Studies* in 1991. The study determined that two quite distinct groups of persons report such experiences: one a stable, normal group, and the other a group showing tendencies toward mental problems, sexual abuse, and proneness to fantasy. A question for future researchers is whether persons in the first group are more likely to describe verifiable, "objective" experiences and persons in the second group unverifiable, "subjective" experiences. Objective experiences would involve, for example, other witnesses and independent evidence, whereas subjective experiences would show no outside witnesses or other evidence.

Popular culture has given the abduction phenomenon growing visibility. This visibility may have produced a model for the paranoid fantasies of troubled persons. In the 1950s, during the height of the Cold War, such persons entertained comparable fantasies of being harassed by Soviet spies. Though Soviet spies certainly did exist, they were not nearly so ubiquitous as reports suggested.

Even skeptical scientists concede that extraterrestrials could well look very much like those abductees describe. A person's imagination could borrow a variety of possible alien forms from science fiction movies, television shows, and novels, even given the popularity of the little gray ET image. It is remarkable that witnesses are so insistent on one particular form: the humanoid.

Not only are abduction reports strikingly similar, but the abductees stubbornly adhere to certain basic details without indulging in the endless elaboration that characterizes folk tales in transmission. Folklorist Bullard finds this the glue that binds such stories. It is not "folk tradition," he says, but "shared experience. A recurrent experience would give similar reports, and they would stay the same as long as the experience remains stable. . . . The folklorist cannot answer what kind of experience the abduction is, but he has sound reasons to say what it is not."

What are UFO abductions? Some of them may be just what they purport to be. Or they could be some extraordinary, heretofore unsuspected mental phenomenon. Possibly some are one, and some are the other. In either case, they form one of the great mysteries of the UFO age. As the scientific investigation of the phenomenon begins, we can expect to learn interesting new truths about ourselves and, perhaps, about the universe in which we—and possibly others—dwell.

CASE STUDIES

HIGHWAY HIJACK
Sandy Larson awoke early the morning of August 26, 1975; so did her 15-year-old daughter, Jackie, and Jackie's boyfriend, Terry O'Leary. Mrs. Larson, who lived in Fargo, North Dakota, was planning to take a real-estate test in Bismarck,

200 miles away. At 4 A.M., 45 miles west of Fargo on Interstate 94, they encountered an unexpected, unimaginable unknown.

First they saw a flash and heard a rumbling sound. Then, in the southern sky, heading east, they saw eight to ten glowing objects with "smoke" around them. One was notably larger than the others, and the witnesses had the impression that in some fashion the other objects had come out of it. The UFOs descended until they were above a grove of trees 20 yards away. Then half of them shot away. The three witnesses suddenly felt an odd sensation, as if they had been frozen or "stuck" for a second or two. Then the UFOs were departing. What was even more weird, Jackie—who had been sitting in the middle of the front seat between Terry and her mother—now sat in the middle of the back seat with no idea how she had gotten there. Moreover, the time now was an hour later.

The following December, Sandy and Jackie separately underwent hypnosis under the direction of University of Wyoming psychologist R. Leo Sprinkle. (Though Terry O'Leary confirmed the sighting and the peculiar feelings associated with it, he declined the offer to explore the incident further.) Jackie remembered being outside the car in a state of paralysis. Her mother told of being floated into the UFO with Terry. A six-foot-tall, robotlike being with glaring eyes put her on a table, rubbed a clear liquid over her, and inserted an instrument up her nose, then performed other medical procedures. Dizzy and nauseated, she felt as if her head would explode. After a period of time she and Terry (whom she did not recall seeing inside the UFO) were returned to their car, and all conscious memory of the incident vanished immediately.

FIRE IN THE ARIZONA SKY

Travis Walton and six other members of a woodcutting crew in the Apache-Sitgreaves National Forest of Arizona were driving home to Snowflake at 6:15 on the evening of November 5, 1975. Along a mountain road they thought they saw, through the trees, a golden, disc-shaped object.

Travis Walton (center) is the chief figure in one of the most spectacular UFO abduction incidents ever recorded.

Fascinated, Walton, 22, jumped out of the truck and ran toward it—only to be hit, the witnesses subsequently testified, by a beam of intense light from the UFO. His body rose a foot into the air, then crashed onto the ground ten feet back.

The others fled in terror. A few minutes later, when they realized the UFO was not pursuing them, they returned to the site, where they expected to find Walton dead. Instead they did not find him at all.

Five days later a weak, confused Walton reappeared. In subsequent interviews—once under hypnotic regression—he reported that he had encountered human and humanoid beings aboard the UFO but had experienced no communication with them. Apparently unconscious during most of this period, he claimed to remember only about two hours of the experience. The episode ended, according to Walton, just before midnight on November 10. He found himself on the highway just outside Heber, Arizona—ten miles from where the UFO first appeared—watching a dark, circular object shooting off into the sky.

The determined efforts of a professional UFO debunker and even some doubting UFO proponents uncovered no real evidence of a hoax. The only negative evidence was Walton's failed polygraph test on November 15. Critics charged that Walton's exhaustion and the polygraph operator's outspoken hostility adversely affected the results, and Walton passed another polygraph test given not long afterward. So did all but one of the crew members. The sole exception gave an "inconclusive" reading. He had not watched the entire event; as it was happening, he cowered in the truck with his head in his hands.

Today Walton and the other crew members still stand by the story. In 1992 Walton passed a fresh set of polygraph tests at the insistence of film-makers who had expressed interest in Walton's 1978 paperback, *The Walton Experience*. The movie, *Fire in the Sky*, was released in March 1993. Its emphasis was on a widespread community suspicion that the young men had murdered Walton and concocted a preposterous yarn to explain his disappearance. Ufologists reacted with dismay to the movie's treatment of Walton's on-board experience. Essentially a Gothic horror sequence, it bore virtually no resemblance to what Walton had reported.

ALIEN IMPLANT?

One commonly reported feature of UFO abduc-tions shows extraterrestrials having an apparent interest in human reproduction. Many abductees also report that alien captors have inserted tiny "implants" into their bodies. The function of these devices is unclear, and proof that they exist at all is ambiguous at best.

An account of a possible implant surfaced in *Nature*, generally regarded as the leading scien-tific journal in the English language. In its September 25, 1986, issue, seven physicians from the genetics department of Churchill Hospital, Headington, England, wrote of a curious discov-ery made during "one of our routine chromosome preparations for prenatal diagnosis following amniocentesis." It looked, they said, "much like a fragmented crossword in appearance."

Bewildered, they solicited suggestions from readers who might recognize it. "Is it a man-made device?" the doctors wondered. "Packing text as binary coded information on the miniature scale would seem advantageous. Or is it a naturally occurring substance? None of the possibilities we have been able to think of would seem to be appropriate for amniotic fluid. . . . We are as intrigued as we are ignorant."

In subsequent issues readers offered various suggestions. The item was variously identified as a diatom skeleton, a "fragment of tubular myelin," an "area of meshwork of the nuclear lamina," and a contaminant originating in the fabrication of semiconductors. In response the senior author of the article, Dr. John Wolstenholme, said that all of these objects are too irregular in appearance to account for the object.

He noted that a number of individuals had written to suggest the object could be of extraterrestrial origin. Wolstenholme rejected that idea but confessed that the precise nature and function of the "mystery object amid the chromosomes" continued to baffle him and his colleagues. It has yet to be explained.

MISSING TIME IN KANSAS

At 11:35 P.M. on November 6, 1989, two St. Louis women were returning home from a conference in Colorado and planned to spend the night in

Goodland, Kansas. Slightly over an hour from there, they noticed that a strange light high in the sky seemed to be keeping pace with them.

At 12:40 A.M. they pulled over to the roadside to get a better look. The object descended to within 200 feet of their vehicle, and as it hovered over a field, a V-shaped cone of multicolored light shone down. The next thing the women knew, their shared sense of excitement had unaccountably turned to irritability and exhaustion.

On driving into Goodland a few minutes later, they were puzzled to find it was 2:30 A.M. A trip that should have taken an hour had taken three. In the morning they stopped at a gas station, where they discovered their three-hour trip had consumed only an hour's worth of fuel.

In the days ahead their anxiety continued. Each of them experienced seemingly inexplicable nose-bleeds. Eventually one sought out hypnotist and psychiatric social worker John Carpenter.

Hypnotized separately, the two women related a virtually identical account of the period of missing time, even though neither discussed her testimony under hypnosis with the other. They reported being floated up into the craft and into a circular room with diffuse, reddish-white light. Slender, humanoid beings with large, slanted eyes, white skin, hairless heads, and four long fingers examined them. The creatures communicated briefly via telepathy and placed a tiny object up one woman's nose. While under hypnosis the other woman recalled a similar implantation but from two weeks earlier.

Carpenter remarked, "At least 40 correlations could be found among descriptions of the creatures, craft interior, behaviors, procedures, and general scenario. . . . No evidence or motivation for a hoax can be found. Both women responded with appropriate emotional relief and amazement at recovering the repressed material."

MUSIC OF THE SPHERES

Connie Cook, a middle-aged divorcee who lives in Peoria, Illinois, sings songs from outer space. Or so she says. Music critics say there is something strange about them. *Rolling Stone* writer David Wild calls her music "otherworldly" in sound, and record producer Jim Van Petten says, "She's using some notes that seem to come from somewhere else. A few don't even translate to music paper."

To play the songs, which a *Wall Street Journal* writer has described as a "dreamy sort of jazz and soft-rock fusion," Cook has had to have her two pianos specially tuned. Lyrically they express New Age sentiments reflecting the peaceful metaphysical philosophy of the people of the Pleiades star system.

Cook says her interactions with space people began with a November 1981 close encounter, when a UFO hovered outside her bedroom window for 90 minutes. When it left, she dreamed she had met its occupants, who told her of their kindly intentions toward terrestrials. Though not a trained musician, she soon began receiving telepathically delivered melodies, which she wrote down. Friends, relatives, and acquaintances describe Cook as sincere and sane.

Earthling-ET collaborations flow from the pen of Danny Endersen, who lives in the Australian state of New South Wales, but except for their exotic content (which includes romantic evocations of otherworldly sexual liaisons), they sound like conventional country-western melodies.

Endersen claims a lifetime of strange experiences, culminating in a 1971 abduction during which he met a seven-foot ET face to face inside a UFO. Subsequently, through an "accelerated learning process," he met a variety of space beings, including some of insect shape and others that looked like eggs. On one occasion aliens healed Endersen's ailing dog. In addition, "they're helping me write some great songs," he says.

A COLD GLARE IN NORWAY

A strange, low-frequency sound was the first indication of something out of the ordinary. It was 10:00 in the evening of March 9, 1992, and two women driving home to Hamar, Norway, not only heard the deep, vibrating sound but felt it in their bodies. Suddenly a dazzlingly bright light loomed up about 50 feet in front of them.

As the driver hit the brakes and the two women adjusted their eyesight, they saw an object inside the light. Hovering less than ten feet above the road, it was triangle-shaped on the bottom, with a rounded cupola of transparent glass on top. The surrounding light vanished abruptly, and the UFO also disappeared from view—to reappear seconds later. Two humanoid beings with large, slanted black eyes sat inside.

The one on the right held something that looked like a steering wheel. Now only a few feet from the object, the women were terrified by the being's cold, malicious glare. The next thing they knew, the beings and the UFO were gone. The witnesses had no memory of how they departed.

Later that evening, as the two discussed the weird encounter in the home of one of them, they heard the sound again just outside the bedroom window. As it grew louder, they were afraid the UFO was going to crash right through the wall. They dived to the floor, even more frightened now than they were during the original encounter. Once again they had no sense of how much time had passed.

Over the next month the two suffered from a variety of physical problems. Both experienced spontaneous nosebleeding and complete loss of energy. One of them lost part of her vision for a week. The women shared an uneasy sense that more had happened to them than they could consciously remember.

REPTOIDS?

Traditionally UFO occupants have been described as humanoid or even humanlike. Until recently little gray-skinned beings with big heads and slanted eyes dominated abduction reporting. Now, according to investigator John Carpenter, reptilian creatures—"hideous, rude, and aggressive"—are showing up in independent accounts.

Carpenter, a hypnotist, says abductees speak of encounters with beings six to eight feet tall. Their green or brown bodies are covered with lizardlike scales, and their faces combine human and serpentine features. Their eyes are catlike, with a vertical slit and gold iris. They have four-fingered claws with brown webbing. They do not communicate with abductees but often treat them roughly. Curiously, Carpenter claims, these presumably extraterrestrial entities are typically encountered inside witnesses' homes, not inside UFOs. Nearly all the reports known so far have come from California and Oklahoma.

Notable abduction researchers Budd Hopkins and David M. Jacobs dismiss such reports as delusional. Many other UFO investigators remain skeptical. Nonetheless sightings of "reptoids" were logged even before ufologists became aware (in the 1960s) of abduction reports.

For example, on the evening of November 8, 1958, Charles Wetzel, driving along the Santa Ana River in Riverside, California, claimed to have seen a bipedal reptilian creature with scales. It bounded across the road in front of his car, then stopped and approached him with a wavering gait, making "gurgling" sounds alternating with screeches. When it began clawing at his windshield, Wetzel accelerated, knocked the creature down, and ran over it. Later police lab tests indicated that something had passed under the car. The claw marks were clearly visible on the windshield. When interviewed again years later, in 1982, Wetzel stuck by his story.

INTO THE
WILD BLUE

Amazing Stories *was the first science-fiction magazine.*

MENACE OF THE UNDEREARTHERS

Before there were little green Martians and tall blond Venusians, there were "deros."

In the 1930s a Pennsylvania man named Richard Sharpe Shaver overheard them speaking through his welding equipment. And the voices—singularly unpleasant ones, obsessed with torture and sexual perversion—would not shut up. Their incessant chatter drove Shaver to desperate acts that landed him in mental hospitals and prisons. While Shaver was serving time in a prison, a woman materialized and whisked him away to a cavern underneath the earth where she and her fellow "teros," though badly outnumbered, battled the dero hordes.

Deros, Shaver explained, were "detrimental robots." The teros were "integrative robots." But neither deros nor teros were actually robots. As with much else, Shaver was vague on the question of why these beings were called robots at all. They were the remnants of a superrace of giants, the Atlans and the Titans, the rest of whom had fled Earth in spaceships 12,000 years ago when the sun began emitting deadly radiation. Those few remaining had retreated to vast caves; during the centuries many degenerated into sadistic idiots (deros) and used the advanced Atlan technology to wreak havoc on the good guys, the teros, who had managed to keep their brains and dignity intact. Other Atlans either stayed on or returned to the earth's surface, adjusted to the new solar radiation, and became our ancestors. To this day deros kidnap and torture surface humans, shoot airplanes out of the sky, and commit other evil acts.

These lurid fantasies enthralled readers of two popular pulp science-fiction magazines, *Amazing Stories* and *Fantastic Adventures,* between 1944 and 1948. Just about anyone else would have pegged Shaver as a complete nutcase and paid no more attention, but he intrigued senior editor Ray Palmer. He snatched Shaver's initial letter out of a wastebasket into which another editor had tossed it with a sneering remark about "crackpots." Soon, as the most intense controversy in the history of science-fiction fandom swirled around him, Palmer vigorously promoted the "Shaver mystery." To many readers it was lunatic nonsense. To others it was the secret of the ages.

Believers who sought "evidence" for dero activity filled the magazines' pages with material gleaned from Charles Fort's writings and from occult lore. Some of this concerned reports of strange ships in the earth's atmosphere. In fact, the June 1947 issue of *Amazing Stories* featured an article on mysterious flying objects that it linked to extraterrestrial visitation. The magazine was on the newsstands when Kenneth Arnold's sighting brought "flying saucers" into world consciousness (see Chapter 1, "In the Beginning").

The Shaver episode, which started just before the UFO age and faded from all but fringe view after its first year, set a standard for tall tales that others would have to scramble to match. Some proved up to the challenge.

As editor of Amazing Stories, *Ray Palmer was the first major commercial exploiter of flying saucers. He promoted some rather exotic theories, notably that saucers were based inside a hollow Earth.*

THE DIRTIEST HOAX OF ALL

From the beginning the urge to spin yarns proved irresistible to some. Like weeds in saucerdom's fertile ground, hoaxes, tall tales, rumors, and other silliness sprouted and spread.

One of the most notorious—and successful—liars, the late Fred L. Crisman, actually bridged the gap between the Shaver mystery and the UFO mystery. Crisman first surfaced in a letter published in the May 1947 issue of *Amazing Stories,* in which he claimed to have shot his way out of a cave full of deros with a submachine gun. Palmer next heard from him the following July. This time Crisman said he had actual physical evidence of a flying saucer.

Palmer passed the story on to Kenneth Arnold, who was investigating reports in the Pacific Northwest. Arnold interviewed Crisman and an associate, Harold Dahl, who identified themselves as harbor patrolmen (they were not). Crisman, who did most of the talking, reported that Dahl had seen doughnut-shaped craft dump piles of slaglike material on the beach of Maury Island in Puget Sound. The next morning a mysterious man in black had threatened Dahl. "I know a great deal more about this experience of yours than you will want to believe," the man had said cryptically.

The two men showed the material to Arnold. In a state of high excitement Arnold contacted an Army Air Force intelligence officer of his acquaintance, Lieutenant Frank M. Brown, who flew up from Hamilton Field in California in the company of another officer. The moment they saw the material, their interest in it evaporated: It was ordinary aluminum. Embarrassed for Arnold, the officers left without telling him their conclusions.

While flying back to Hamilton, their B-25 caught fire and crashed, killing both officers. Though Crisman and Dahl subsequently confessed to other Air Force investigators that they had made up the story, the legend would live on for decades afterward. Some writers—including Arnold and Palmer, who wrote a book about the case—hinted that the officers died because they knew too much. But to Captain Edward Ruppelt of Project Blue Book, the Maury Island incident was the "dirtiest hoax in UFO history."

Years later Crisman's name would reemerge in another contentious context. New Orleans district attorney Jim Garrison, while investigating what he believed to be a high-level conspiracy to murder President John F. Kennedy, called Crisman in December 1968 to testify before a grand jury. Some early assassination-conspiracy theorists would identify Crisman (falsely) as one of the three mysterious "hoboes" arrested and photographed shortly after the shooting in Dallas.

Before his death Crisman was peddling a new, improved, UFO-less version of the Maury Island story. He now claimed that the "truth" involved, not flying doughnuts dropping slag, but something even more dangerous: illegal dumping by military aircraft of radioactive waste into the harbor. Though this tale was no less tall than his earlier one, it has already entered UFO literature as the "solution" to the Maury Island "mystery."

FROM OUTER SPACE TO YOUR WALLET

Consider the case of George Adamski. Born in Poland in 1891, Adamski came to America in his infancy. He received a spotty education and developed an early interest in occultism. By the 1930s Adamski had established a niche as a low-rent guru in southern California's mystical scene. He founded the Royal Order of Tibet, whose teachings drew on his psychic channelings from "Tibetan masters." In the late 1940s "Professor" Adamski produced pictures of what he said were spaceships he had photographed through his telescope.

*New York radio and television personality
Long John Nebel (right) provided George Adamski (left)
with a forum to promote his books and
photographs, though Nebel did not
hide his personal skepticism.*

The pictures attracted wide attention. But the events that began on November 20, 1952, would make Adamski a saucer immortal. Responding to channeled directions from extraterrestrials (who had replaced the Tibetan masters, though their messages were identical), Adamski and six fellow occult seekers headed out for the desert. Near Desert Center, California, he separated from the others and met a landed spaceship. Its pilot was a friendly fellow named Orthon, a handsome, blond-haired Venusian.

Serious UFO investigators scoffed, but other people all over the world believed, even as Adamski's tales grew ever more outrageous. Adamski's 1955 book, *Inside the Space Ships*, recounted his adventures with Venusians, Martians, and Saturnians, who had come to Earth out of concern for humanity's self-destructive ways. These "Space Brothers," as Adamski and his disciples called them, proved a long-winded lot, fond of platitudes and full of tedious metaphysical blather.

In Adamski's wake other "contactees" emerged to spread the interplanetary gospel and to count the take at gatherings of the faithful. The principal gathering was held every summer at Giant Rock, near Twentynine Palms, California. The driving force behind the Giant Rock gatherings was George Van Tassel, who had established psychic contact with extraterrestrial starships ("ventlas") in early 1952. A few months later he rushed into print the first modern contactee book, the misleadingly titled *I Rode a Flying Saucer!* The following year, Van Tassel would get to do just that when his pal Solganda invited him inside a spaceship for a quick spin.

At the space people's direction, Van Tassel established the College of Universal Wisdom and solicited donations for the construction of the Integratron, a rejuvenation machine. When completed, Van Tassel told his supporters, it would handle as many as 10,000 persons a day. People would emerge looking no younger, but their cells would be recharged. Untold tens of thousands of dollars later, the Integratron languished unfinished in February 1978 when Van Tassel died—from the ravages of old age.

Reincarnated Saturnian and space communicant Howard Menger held forth from his farm in New Jersey, where followers would come to witness—well, something. Followers would see lights and even figures but always in the dark and never up close. Once, when Menger led a follower into a dark building to speak with a spacewoman, a sliver of light happened to fall on the face of the "extraterrestrial." It was, the follower could not help noticing, identical to the face of a young blond woman who happened to be one of Menger's closest associates.

After releasing a book, *From Outer Space to You* (1959), and a record album, *Music from Another Planet,* Menger would virtually recant his story, vaguely muttering about a CIA experiment. In the late 1980s he withdrew his recantation and marketed a new book detailing his latest cosmic adventures.

In the late 1950s, followers called New Jersey contactee Howard Menger the "East Coast Adamski" because his story was so similar to his California counterpart's.

Most contactees have managed to stay out of legal trouble, though law-enforcement and other official agencies look into their activities from time to time. Reinhold Schmidt was not so lucky. In the course of contacts with German-speaking Saturnians, Schmidt's space friends showed him secret stores of quartz crystals in the mountains of California. Armed with this information and a gift for (so the prosecutor charged) "loving talk," he persuaded several elderly women to invest their money in a crystal-mining venture. The money went, however, into his own pocket. He was put on trial for grand theft and from there went to jail.

Still, not all contactees are con artists, by any means. In 1962 Gloria Lee, who chronicled her psychic contacts with "J.W." of Jupiter in *Why We Are Here* (1959), starved to death in a Washington motel room after a two-month fast for peace ordered by her space friends. In 1954, in the face of massive press ridicule, followers of Dorothy Martin, who communicated with extraterrestrials through automatic writing, quit jobs and cut all other ties as they awaited a prophesied landing of a flying saucer that would pick them up just before geological upheavals caused massive destruction.

The charlatan contactees typically claim physical encounters, nearly always have photographs and other artifacts to prove it (in one especially brazen instance, packets of hair from a Venusian dog), and in general behave more like profiteers than prophets. The psychic contactees, on the other hand, tend to be quiet, unflamboyant, and almost painfully sincere. They can best be described as Space Age religious visionaries. In another century their messages would have been from gods or angels or spirits. These messages, generally inane and rarely profound, are manifestly not from true extraterrestrials. Psychologists who have studied contactees believe these individuals are not crazy, just unusually imaginative; their communicators come from inner, not outer, space, through a nonpathological form of multiple-personality disorder.

Though only a few professional contactees of the 1950s are still alive or active today, the contactee movement is as big and vibrant as ever. This is due in part to the efforts of a Laramie, Wyoming, psychologist, R. Leo Sprinkle, who sponsors an annual summer conference on the University of Wyoming campus. Those attending are mostly individuals convinced that the Galactic Federation—a sort of extraterrestrial United Nations—has placed them on Earth to spread the cosmic gospel. In a sense these conferences function as revival meetings in which the faith is renewed even as the larger world continues to scoff.

THE DARK SIDE

While contactees offer a rosy picture of the UFO phenomenon, other, darker visions have obsessed some saucer enthusiasts. In fact, even contactees agree that all is not well. Sinister forces oppose the Space Brothers' benevolent mission. Some of these are extraterrestrial and others terrestrial, and they work together to thwart the emergence of the truth.

Among the early victims of this evil "Silence Group" was Albert K. Bender of Bridgeport, Connecticut. In 1952 Bender formed the International Flying Saucer Bureau (IFSB), which met with immediate success, but he shut it down the next year under mysterious circumstances. In due course Bender confided that three men in black had imparted to him the terrifying answer to the UFO mystery and turned his life into a nightmare. He would say no more. Three years later an IFSB associate, Gray Barker, wrote a book about the episode. The title perfectly captured the paranoia abroad in UFO-land: *They Knew Too Much About Flying Saucers*.

Through the "Bender mystery" the legend of the "men in black" (MIB) came into the world—even though, as Barker observed, a man in black had played a villainous role in the Maury Island incident. According to Barker, the MIB were ranging as far afield as Australia and New Zealand, scaring still more UFO buffs into silence.

By the late 1980s MIB tales had become so prevalent that the distinguished *Journal of American Folklore* took note of them in a long article. Just who the MIB were remained unclear. To saucerians enamored with conspiracy theories, they were enforcers for the Silence Group, associated with international banking interests that sought to stifle the technological advances and moral reforms the Space Brothers wanted to bestow on earthlings. To others, they were alien beings—perhaps, some speculated, Shaver's deros. In 1962 Bender came down on the side of the alien school. Breaking his nine-year silence in *Flying Saucers and the Three Men,* which he insisted was not a science-fiction novel, Bender revealed that the men in black who drove him out of ufology were monsters from the planet Kazik. Even Barker, the book's publisher and a relentless Bender promoter, remarked privately and out of customers' hearing, that maybe it had all been a "dream."

Fear of the MIB was generated in part by worries about the possibly hostile motives of UFOs. A popular early book, *Flying Saucers on the Attack,* by Harold T. Wilkins (1954), fretted that a "Cosmic General Staff" could even now be plotting a real-life war of the worlds. But compared with demonologist-ufologist John A. Keel, author of *UFOs: Operation Trojan Horse* (1970) and other writings, Wilkins sounded like an optimist. In Keel's vision, UFO intelligences are not simply extraterrestrials but "ultraterrestrials"—entities from unimaginable other dimensions of reality. Worse, they do not like us at all. Human beings, Keel thunders, are "like ants, trying to view reality with very limited perceptive equipment. . . . We are biochemical robots helplessly controlled by forces that can scramble our brains, destroy our memories and use us in any way they see fit. They have been doing it to us forever."

In recent years new and even wilder strains of paranoia have sprouted along ufology fringes. Inspiration comes not just from UFO rumors but from conspiracy theories associated with the far right end of the political spectrum. The two major figures in what has been called the "dark-side movement" are John Lear, a pilot who once flew aircraft for a CIA-linked company, and Milton William Cooper, a retired Navy petty officer.

According to dark-siders, a ruthless "secret government" controls the world. Among other nefarious activities, it runs the international drug trade and has unleashed AIDS and other deadly diseases as population-reducing measures. Its ultimate goal is to turn Earth and surrounding planets into slave-labor camps. For some time this secret government has been in contact with alien races, allowing the aliens to abduct human beings in exchange for advanced alien technology.

The aliens, known as the "grays" (because of their gray skin color), do more than abduct human beings. They mutilate and eat them as well, using the body parts to rejuvenate themselves. The secret government and the aliens labor together in vast underground bases in New Mexico and Nevada, where they collect human and animal organs, drop them into a chemical soup, and manufacture soulless android creatures. These androids, who are then unleashed to do dirty work for the government/alien conspiracy, are best known to the rest of us as the men in black.

With each retelling, with the appearance of each new and expensive book, video, or tape, the dark-side story gets crazier. In one version the conspirators travel into the future to observe the emergence of the anti-Christ in the 1990s, World War III in 1999, and the Second Coming of Christ in 2011. George Bush oversaw the world's drug traffic. The secret government has maintained bases on Mars since the early 1960s. The conspirators employ drugs and hypnosis to turn mentally unstable individuals into mass murderers of schoolchildren and other innocents; the purpose is to spur antigun sentiment, resulting in gun control legislation. Thus, Americans will be disarmed and defenseless when the secret government's storm troopers round them up and herd them into concentration camps.

A small army of fervent believers all around the world has embraced these monstrous yarns, for which—no rational reader will be surprised to learn—not a shred of supporting evidence exists. The true sources of these lurid tales are not hard to find: They are a hodgepodge of elements patched together from saucer folklore, extremist political literature, and a 1977 British mock-documentary, *Alternative 3.* The purpose of this show was to satirize popular credulity and paranoia. Unfortunately, some remain convinced the show was sober fact—which, ironically, produced fresh varieties of mass gullibility and fear.

THE ALIENS ON THE PLAINS

Before he died in 1969, U.S. Soil Conservation Service engineer Grady "Barney" Barnett confided a fantastic story to friends. Some years earlier, he said, while working in the New Mexico desert, he had come upon a crashed flying saucer and its dead humanoid occupants. A team of archaeologists was already at the site when he arrived. In short order, military personnel showed up and swore the civilian witnesses to silence.

Barnett's friends, who took him seriously, recalled that he had placed the incident on the Plains of San Agustin, but they did not know the year. In the late 1970s investigators began looking into another UFO-crash story, this one (now known as the Roswell incident) set in Lincoln County, New Mexico, in early July 1947 (see Chapter 6, "The Ultimate Secret"). The investigators concluded that the event at San Agustin, 150 miles to the west, had taken place at the same time. With Barnett no longer available for direct interview, the investigators had to rely on friends' memories of his testimony from years before. Barnett's story appeared in print for the first time in Charles Berlitz and William L. Moore's *The Roswell Incident* (1980).

Years of diligent searching failed to uncover any living witnesses to a crash at San Agustin in 1947 or at any other time. Then in January 1990, after NBC's popular television series *Unsolved Mysteries* aired a segment on the alleged incident, the long-sought witness stepped forward.

Gerald Anderson of Springfield, Missouri, claimed that when he was five years old, he had gone on a rock-hunting expedition with his family to the Plains on July 5, 1947. There, in the company of a man who looked like President Truman (a fair description of Barnett) and an archaeological team led by a "Dr. Buskirk," the Andersons saw the saucer and the aliens—one of them still alive—and were threatened by soldiers.

Not everyone who had been looking for San Agustin witnesses was taken with Anderson's account. Kevin D. Randle and Donald R. Schmitt, who were researching the Roswell incident for a book (published in 1991 as *UFO Crash at Roswell*), noted discrepancies in the story, found his memory suspiciously detailed, and soon judged Anderson untrustworthy. But Stanton T. Friedman and Don Berliner championed Anderson, using his story as evidence that two extraterrestrial spacecraft collided, one crashing in Lincoln

Physicist and lecturer Stanton T. Friedman—shown here holding a portrait of a humanoid based on the recollections of a New Hampshire man who claims he was abducted in 1971—championed Anderson.

County, the other on the Plains. Anderson figures prominently in their 1992 book *Crash at Corona*.

For a time Anderson became something of a UFO celebrity. His hometown newspaper recounted his story in a long, uncritical article, and he was featured in television programs, videos, and magazine articles. He spoke at gatherings of UFO buffs. He also passed a polygraph test arranged by his supporters. Even so, as critics observed, his story changed in the telling; the date and even the location varied from account to account.

Moreover, all other family members were dead or untraceable. When skeptics remarked on this convenient absence of corroborating testimony, Anderson produced a 1947 diary said to have been written by his late Uncle Ted, another alleged witness, and a recent letter supposedly from Ted's daughter, who was purportedly being hidden by the Roman Catholic Church. Both documents proved bogus. Anderson's ex-wife testified to his long fascination with UFOs and science fiction as well as his penchant for the spinning of tall tales. Investigator Thomas J. Carey learned that "Dr. Buskirk," still alive, was Anderson's anthropology teacher at Albuquerque High School in 1957. Anderson's sketch of him showed Buskirk as he looked then, not as he looked in 1947. That year Buskirk had been in Arizona, doing anthropological work with the Apache tribe, and had never set foot in New Mexico. He had even published a book on his research that summer.

Eventually, after Anderson confessed to forging a document that critics had long characterized as phony, most of Anderson's supporters gave up on him. His few remaining defenders, notably Friedman, hint that Anderson has been the victim of a frame-up, possibly engineered by agents of the UFO cover-up.

Sometimes a hoax—even a not very good hoax—is hard to kill. We probably have not heard the last of Gerald Anderson.

CASE STUDIES

CROP CIRCLES

In August 1980 farmer John Scull of Bratton, Wiltshire, England, found seemingly inexplicable flattened circles in his oats field. Every summer thereafter, southern England has been inundated with crop circles.

Science writer Dennis Stacy, who has investigated and written extensively on the crop circle phenomenon, says, "In a true crop circle the individual stalks are laid over without breaking or other obvious overt damage, even in the case of notoriously fragile plants. . . . That is, the plant stalks appear to go limp, almost as if they had been steamed or rendered elastic, before returning to their previous state. An encircling ring may run counterclockwise to the central circle and vice versa, or both may be laid in the same direction."

Over time the formations grew from simple circles to complex "pictograms" depicting some-

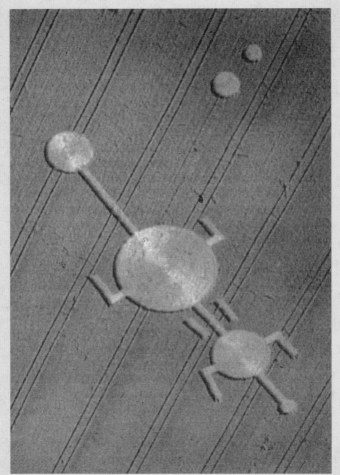

Some of the crop circles have been of elaborate design. This arrangement, seen in Goodworth Clatford, Hampshire, in 1991, has been said to resemble a "spacelike" person.

times complex geometric shapes. Some scientists, most notably meteorologist Terence Meaden of the Tornado and Storm Research Organization, have theorized that the "simple" circles are formed by an unusual natural process that he calls a "plasma vortex": the energetic breakdown of a standing, electrically charged whirlwind. As it falls apart, it descends in a burst of violent wind that cuts out circles on the ground beneath it. Some other scientists doubt that such an effect is physically possible, but none has proposed a plausible counterexplanation.

Perhaps the best argument against a natural explanation is that no pre-1980 records of clear, unambiguous crop circles seem to exist. Various candidates, some going back several centuries, have been proposed, but none is entirely convincing. It is extremely unlikely that a meteorological effect could have sprung into creation only recently.

In September 1991 two British sexagenarians, Doug Bowers and David Chorley, confessed to having hoaxed as many as 250 of the circles. Other hoaxes have been perpetrated by university students and other pranksters. By 1992 hoax circles had become an epidemic. Paul Fuller, editor of the magazine *Crop Watcher,* observed that even the boldest enthusiasts were not prepared to "stick their necks out and claim that [just] a single [1992] British circle qualified as 'genuine'. . . . [T]he awful truth has dawned on cerealogists [crop-circle researchers] everywhere—that most modern crop circles really are manmade hoaxes

137

and that if there ever was a 'genuine' phenomenon in the first place it has been utterly swamped by a smoke screen of wishful thinking and media inspired mythology."

THE SAUCERS OF S-4

On November 11 and 13, 1989, viewers of KLAS-TV in Las Vegas, Nevada, heard an incredible story from news reporter George Knapp: A scientist had come forth to reveal that the U.S. government possesses the remains of extraterrestrial vehicles. From these vehicles have come extraordinary technological breakthroughs.

The scientist, Robert Scott Lazar, said he had worked in the S-4 section of Area 51, a corner of the Nevada Test Site. There he had read documents indicating the existence of ongoing research on an "antigravity reactor" for use in propulsion systems. He was astonished, he said, but he was even more shocked to be shown nine flying discs "of extraterrestrial origin" stored in a hangar. As part of the gravity-harnessing propulsion, the craft used an element, 115, unknown on Earth, because it is "impossible to synthesize an element that heavy here on Earth. . . . The substance has to come from a place where super-heavy elements could have been produced naturally." From the recovered craft the U.S. government had collected some 500 pounds of the stuff.

Adding apparent credibility to Lazar's testimony were persistent reports (chronicled even in the respected *Aviation Week & Space*

Technology) of bizarre lights over the test site—craft maneuvering in ways beyond the capacity of known aviation technology. These reports are almost certainly genuine.

Lazar's tales, on the other hand, are almost certainly bogus. Investigations raised serious questions about his reliability. His claims about his education and employment could not be verified, and his character proved to be questionable. In 1990 he was arrested for involvement with the operation of a Nevada brothel.

MON-KA OF MARS

Mon-Ka is many people's favorite Martian. According to one chronicler, he "has a wisdom that is light-years beyond the most intelligent person on our planet."

Mon-Ka's first communication with earthlings was in April 1956 at the Giant Rock Spacecraft Convention in southern California, when contactee Dick Miller played recordings that he said had mysteriously appeared on tapes in sealed cans. On the tapes Mon-Ka asked a favor and made a promise: "On the evening of November 7, of this your year 1956, at 10:30 P.M. your local time, we request that one of your communications stations remove its carrier signal from the air for two minutes. At that time we will speak from our craft, which will be stationed at an altitude of 10,000 feet over your great city of Los Angeles."

In September Miller went to London and played the tapes for impressionable British saucer fans.

The Associated Press's tongue-in-cheek treatment afforded the story international attention. A subsequent Los Angeles *Mirror-News* account revealed that Miller had once faked a radio communication from a saucer in his native Detroit. Nonetheless, southern California succumbed to Mon-Ka mania. Two mass rallies were held in Los Angeles in late October, and organizer Gabriel Green enthusiastically talked up the Martian on Art Linkletter's popular *House Party* television show.

When the evening of November 7 rolled around, the faithful climbed to the rooftops and scanned the skies. As publicity gimmicks two radio stations went off the air at the appointed hour, and a television station sent out a plane to look for the Martian spaceship. Mon-Ka did not show up.

He was not, however, gone for good. Since then Mon-Ka has channeled psychic communications to numerous contactees. Today he is beloved as a tireless (and garrulous) "soldier for the cause of peace," in the words of one admirer.

THE STATE DEPARTMENT AND THE VENUSIANS

George Adamski was one of the most famous—or notorious—figures on the flying saucer scene from 1952 until his death in 1965. In books and lectures he recounted his meetings with friendly Venusians, Martians, and Saturnians. He also claimed that high government officials—themselves in contact with "Space Brothers"—secretly knew he was telling the truth.

Nonetheless, Adamski was shocked one day in December 1957 to receive a letter written on U.S. State Department stationery with a stamped department seal and a Washington, D.C., postmark. Signed by "R. E. Straith, Cultural Exchange Committee," it stated, "The Department has on file a great deal of confirmatory evidence bearing out your own claims. . . . While certainly the Department cannot publicly confirm your experiences, it can, I believe, with propriety, encourage your work."

The Straith letter electrified Adamski's followers. They charged the department with covering up the truth when the department denied, as it did repeatedly, that it knew anything of an "R. E. Straith" or a "Cultural Exchange Committee." All the while Straith proved elusive; despite repeated efforts, Adamski's supporters could not find him. Undaunted, they concluded that his committee must be so highly classified that the government would never admit to its existence.

Ufologists skeptical of Adamski's claims were sure the letter was a forgery—perhaps, as analyst Lonzo Dove suspected, typed by Gray Barker, a saucer publisher and practical joker. When Dove submitted an article on the subject to *Saucer News* editor Jim Moseley, Moseley rejected it on the grounds that Dove had not proved his case. But years later, after Barker's death in December 1984, Moseley confessed that he and Barker had written the letter on official stationery provided by a friend of Barker's, a young man who had a relative who was high in the government.

SPACESHIP CRASH IN 1884?

On June 6, 1884, as a band of cowboys rounded up cattle in remote Dundy County, Nebraska, a blazing object streaked out of the sky and crashed some distance from them, leaving (according to a contemporary newspaper account) "fragments of cog-wheels and other pieces of machinery . . . glowing with heat so intense as to scorch the grass for a long distance around each fragment." The light was so intense that it blinded one of the witnesses.

This incredible event was recorded two days later in Lincoln's *Daily State Journal,* which printed a dispatch from Benkelman, Nebraska, by an anonymous correspondent. The correspondent wrote that prominent local citizens had gone to the site, where the metal now had cooled. He reported, "The aerolite, or whatever it is, seems to be about 50 or 60 feet long, cylindrical, and about ten or 12 feet in diameter." A *Daily State Journal* editor remarked that this must have been an "air vessel belonging originally to some other planet."

But on June 10 an anticlimactic dispatch came from Benkelman. In a heavy rainstorm the remains had "melted, dissolved by the water like a spoonful of salt." The obvious message: Take the story with a grain of sodium chloride. The *State Journal,* red-faced, dropped it then and there.

In the 1960s a copy of the first newspaper article resurfaced, and reporters, historians, and ufologists rushed to Dundy County. Lifelong residents of the area assured them no such thing had ever happened. Later, even after the telltale follow-up dispatch was uncovered, one humorless author theorized that the "storm was artificially created so that a UFO concealed within the clouds could retrieve the wreckage of the crashed UFO."

SALVATION FROM SPACE

Every summer contactees—people who believe they have communicated with godlike space people—flock to the Rocky Mountain Conference on UFO Investigation, held on the University of Wyoming campus in Laramie. All these people have remarkable stories to tell: stories of personal transformation that sound like classic religious experiences in Space Age guise.

One of the stories is told by Merry Lynn Noble, who by her own admission was once "one of the leading call girls in the western United States." She was also an alcohol and drug addict seeking to change her life through spiritual studies. In February 1982, exhausted and depressed, she visited her parents in Montana. One evening, as they were driving in the country, a flying saucer appeared, bathing the car in light.

Noble's parents, who "were just frozen there," seemed unaware of the UFO's presence. Meanwhile, Merry Lynn in her astral body was being drawn into the craft, where she felt "absolute ecstasy, total peace, womblike warmth. . . . 'I'm so glad to leave that body,' I thought." She communicated telepathically with a "presence" who gave her a "new soul, with new energy, new humility." The next thing she knew, she was jolted back into her physical body.

From that moment her life began to change for the better. She found a good job and joined Alcoholics Anonymous. Her psychic contact with the extraterrestrial she met aboard the saucer continues, and she has written an unpublished autobiography, *Sex, God and UFOs*.

TO THE MOON, OTIS!

To hear him tell it, Otis T. Carr was the smartest man since Isaac Newton, Albert Einstein, and Nikola Tesla. Not only that, but Tesla, the great electrical genius and contemporary of Thomas Edison, had confided some of his deepest secrets to Carr when the latter worked as a young hotel clerk in New York City in the 1920s.

In the mid-1950s, with Tesla long gone, Carr was ready to tell the world and collect the rewards. He founded OTC Enterprises, hired a fast-talking business manager named Norman Colton, and set out to secure funding for a "fourth-dimensional space vehicle" powered by a "revolutionary Utron Electric Accumulator." The saucer-shaped OTC-X1 would undergo its first flight in April 1959 and the following December would go on all the way to the moon.

Carr and Colton secured hundreds of thousands of dollars from wealthy investors and contactee-oriented saucer fans, including Warren Goetz, who claimed to be an actual space person, having materialized as a baby in his (Earth) mother's arms while a saucer hovered overhead.

Otis T. Carr holds his "electric accumulator."

Another associate, Margaret Storm, wrote a biography of Tesla, who turns out to have been a Venusian. To skeptics Carr was a shameless spouter of double-talk and baffle-gab. As one observer put it, "For all most people know, he might well be a great scientist. After all, he is completely unintelligible, isn't he?"

On Sunday, April 19, 1959, while crowds gathered at an amusement park in Oklahoma City to watch the OTC-X1's maiden flight, Carr suddenly contracted a mysterious illness and had to be hospitalized. He mumbled something about a "mercury leak," but burly guards kept reporters who wanted to check for themselves out of the plant where the craft supposedly was being constructed.

The OTC-X1 never went to the moon, but Carr went to prison for selling stock illegally. He died penniless years later in a Pittsburgh slum. Colton, who had skipped out of Oklahoma a step ahead of the authorities, formed the Millennium Agency, which sold stock in machines "operated entirely by environmental gravitic forces." They never flew, either.

"ETHER SHIPS"?

In a 1950 monograph California occultist N. Meade Layne proposed that UFOs and their occupants come here not from other planets but from another order of reality. Layne called this place Etheria and declared that it surrounds us yet is usually invisible. Psychically inclined individuals are most attuned to it, but some of its manifestations, such as flying saucers, can be seen by anyone. The saucers can materialize and dematerialize; at certain stages they are "jellylike," enabling them to "change in shape and apparent size." Layne's theory, at least in general, would survive to be championed by ufologists (most notably John Keel and Jacques Vallée).

Investigators had doubted the authenticity of this UFO photo, chiefly because of the artificial-looking "shadow." Later, the Venezuelan airline pilot who took the picture confessed that the "UFO" was merely a button.

DEVILS OR ANGELS?

To fundamentalist Christians, UFO beings are either demons or angels. Most fundamentalist writers favor the first interpretation. Kelly L. Segraves, for example, holds that these beings are "fallen angels and followers of Satan" who seek to lead us into "depravity and rejection of God." Clifford Wilson believes Satan's agents have abducted human beings into UFOs and turned them into agents as part of "some great super-plan of a spiritual counterattack to reach its culmination in Armageddon." But to the most famous evangelist of all, Billy Graham, UFOs are "astonishingly angellike." He has said he believes they are here to prepare us for Jesus' return.

MASTER OF THE UNIVERSE

As galactic heavyweights go, few tip the scales as impressively as Ashtar, commander of the 24 million extraterrestrials involved in the Earth project. According to one of his Earth friends, Ashtar is sponsored by "Lord Michael and the Great Central Sun government of this galaxy. . . . Second only to the Beloved Commander Jesus-Sananda in responsibility for the airborne division of the Brotherhood of Light." Ashtar beams his channeled messages from a colossal starship, or space station, that entered the solar system on July 18, 1952.

The first to hear from him was California contactee George Van Tassel, but since then dozens, and possibly hundreds, all over the world have heard from him and communicated his sermons. Asked what he looks like, Ashtar replies modestly, "I am seven feet tall in height, with blue eyes and a nearly white complexion. I am fast of movement and considered to be an understanding and compassionate teacher."

NAKED SPACE PEOPLE?

Two outlandish yet similar tales told half a century apart seem to indicate that extraterrestrial beings may at times appear nude. On April 19,

1897, the *St. Louis Post-Dispatch* printed a letter from one W. H. Hopkins. Three days earlier, near Springfield, Missouri, Hopkins encountered a beautiful nude woman standing outside a landed "airship." As he approached, a similarly unclad man stepped up to protect her. Though neither being spoke English, Hopkins convinced them of his peaceful intentions. Asked where they came from, they "pointed upwards, pronouncing a word which sounded like Mars." On March 28, 1950, Samuel Eaton Thompson reportedly met up with nude Venusian men, women, and children in a forest outside Mineral, Washington. Friendly but childlike, they spoke "uneducated" English. Whereas Hopkins' Martians were sweating in the spring temperatures, Thompson's Venusians were cold because of the respective distances of the two planets from the sun.

ARE YOU A STAR PERSON?

Contactee chronicler Brad Steiger says you may be a Star Person if you are physically attractive, have a magnetic personality, require little sleep, hear unusually well, work in the healing or teaching profession, and harbor the suspicion that this world is not your home. Steiger discovered this while working on a book on space channeling. When he announced his discovery in the May 1, 1979, issue of the *National Enquirer,* he was inundated with letters from people who recognized themselves.

According to Steiger, there are four kinds of Star People—Refugees, Utopians, Energy Essences, and reincarnated ETs. All these Star People have been placed on Earth to prepare it for the great changes that will come in the wake of a series of worldwide disasters. These disasters will precede mass landings by the Star People's relatives in other worlds. In his 1981 book, *The Star People,* Steiger predicted such cataclysms as a pole shift and worldwide famine in 1982, World War III in the mid-1980s, and Armageddon in 1989.

A MAN IN BLACK

In 1987, writing in the respected *Journal of American Folklore,* Peter M. Rojcewicz examined "folk concepts and beliefs in 'other worlds'" as they related to "men in black" (MIB) legends. One classic tale of an MIB involved a man with the pseudonym "Michael Elliot." One afternoon, as Elliot sat in a university library immersed in UFO literature, a thin, dark-featured man approached him. Speaking in a slight accent, the man asked Elliot what he was reading about. Flying saucers, Elliot replied, adding that he had no particular interest in their reality or unreality, just in the stories told about them. The stranger shouted, "Flying saucers are the most important fact of the century, and you're not interested!?" Then the man stood up "as if mechanically lifted"; spoke gently, "Go well in your purpose"; and departed. When Elliot went to follow the man, he found the library eerily deserted. A year or two after his article appeared, Rojcewicz confessed that he was "Michael Elliot."

Albert Bender drew this sketch of one of the mysterious men in black (see page 128). The MIB provide endless fodder for UFO buffs of a paranoid disposition.

THE ULTIMATE
SECRET

Army Finds Air Saucer
On Ranch in New Mexico

**Disk Goes
To High
Officers**

*Posted Up
Late Week*

ROSWELL, N.M.—(AP)—
The Army Air Forces
today announced a Flying
Saucer had been found on a
lonely New Mexico ranch in its
Army Air Forces...

'Flying disc'
turns up as
just hot air

Fort Worth, Tex., July 9 (AP) — An examination by the Army revealed last night a mysterious object found on a lonely New Mexico ranch was a harmless high-altitude weather balloon — not a grounded flying disc.

Army Knocks Down Disk—
IT'S A WEATHER BALLOON

Device Is Only
A Wind Target

**Object Found in N. Mexico
Identified at Fort Worth**

The official explanation of the Roswell incident.

RUMORS OF THE INCREDIBLE

The stories began to circulate in the late 1940s. They were so fantastic that even those willing to seriously consider the possibility of extraterrestrial visitation responded with incredulity.

In fact, no more than a couple of weeks after Kenneth Arnold's sighting ushered in the UFO age, the first such story hit the press. One morning in early July 1947, Mac Brazel, foreman of a ranch located near tiny Corona, New Mexico, rode out on horseback to move sheep from one field to another. Accompanying him was a young neighbor boy, Timothy D. Proctor. As they rode, they came upon strange debris—various-size chunks of metallic material—running from one hilltop, down an arroyo, up another hill, and running down the other side. To all appearances some kind of aircraft had exploded. In fact Brazel had heard something that sounded like an explosion the night before, but because it happened during a rainstorm (though it was different from thunder), he had not stepped out to look into the cause. Brazel picked up some of the pieces. He had never seen anything like them before. They were extremely light and very tough.

On the afternoon of July 8, 1947, the Roswell, New Mexico, *Daily Record* startled the nation with a report of a flying saucer crash northwest of Roswell and of the recovery of the wreckage by a party from the local Army Air Force base. Soon, however, the Air Force assured reporters that it had all been a silly mistake: The material was from a downed balloon.

Though this particular incident was quickly forgotten, rumors of recovered saucers and, in addition, the bodies of their alien occupants, became a staple of popular culture—and con games. In 1949 *Variety* columnist Frank Scully wrote that a "government scientist" and a Texas oilman had told him of three crashes in the Southwest. The following year Scully expanded these claims into a full-length, best-selling book, *Behind the Flying Saucers*, which claimed that the occupants of these vehicles were humanlike Venusians dressed in the "style of 1890." But two years later *True* magazine revealed in a scathing exposé that Scully's sources were two veteran confidence men, Silas Newton and Leo GeBauer. Newton and GeBauer were posing respectively as an oilman and a magnetics scientist in an attempt to set up a swindle involving oil-detection devices tied to extraterrestrial technology.

To serious ufologists, including those who suspected the government wasn't telling everything it knew about UFOs, crash stories were farfetched yarns of "little men in pickle jars." A person with such a story got a chilly reception when he or she passed it on to anyone but fringe ufologists. In 1952 Ed J. Sullivan of the Los Angeles–based Civilian Saucer Investigators wrote that such tales "are damned for the simple reason, that after years of circulation, not one soul has come forward with a single concrete fact to support the assertions. . . . We ask you to beware of the man who tells you that his friend knows the man with the pickle jar. There is good reason why he effects [*sic*] such an air of mystery, why he has been 'sworn to secrecy'—because he can't produce the friend—or the pickle jar."

Rumors persisted nonetheless. In 1954, after President Dwight Eisenhower dropped out of sight while visiting California (sparking a press-wire report that he had died), it was alleged that he had taken a secret trip to Edwards Air Force Base (AFB) to view alien remains—or, as another version had it, to confer with living aliens. A soldier with the Air Force confided that in 1948 he and other soldiers were dispatched to a New Mexico site to dismantle a nearly intact craft, from which an earlier party had removed the bodies of little men. In Europe it was said that the Norwegian military found a saucer on the remote North Atlantic island of Spitsbergen, or maybe it was the German military and the island was Heligoland. On May 23, 1955, newspaper columnist Dorothy Kilgallen wrote, "British scientists and airmen after examining the wreckage of one mysterious flying ship are convinced that these strange aerial objects are not optical illusions or Soviet inventions but are actual flying saucers which originate on another planet."

Over Chesapeake Bay on the evening of July 14, 1952, the pilot and copilot of a Pan American DC-3 had a much-publicized encounter with eight plate-shaped UFOs that maneuvered in formation near their aircraft. The next morning, as they waited to be interviewed separately by Air Force officers, the two agreed to ask about the crash rumors. Subsequently, the copilot, William

Fortenberry, raised the question, and one of the interrogators replied, "Yes, it is true." Pilot William Nash forgot to ask until afterward, when he and Fortenberry met together with the officers. Nash recalled, "They all opened their mouths to answer the question, whereupon Major [John H.] Sharpe looked at them, not me, and said very quickly, 'NO!' It appeared as if he were telling them to shut up rather than addressing the answer to me." Later Nash met a New York radio newsperson who claimed the Air Force had briefed him and two other reporters (one from *Life* magazine) about its recovery of a crashed UFO.

TESTIMONY OF A SCIENTIST

A remarkable interview occurred in Washington, D.C., on September 15, 1950, but the content did not leak out until the early 1980s, when Canadian ufologist Arthur Bray found a memo by one of the participants, radio engineer Wilbert B. Smith of Canada's Department of Transport. The memo described a conversation with physicist Robert I. Sarbacher, a consultant with the U.S. Department of Defense Research and Development Board (RDB), at one of the regular meetings Sarbacher and other government scientists conducted with their Canadian counterparts on national defense matters.

Asked about the crash rumors, Sarbacher said they were "substantially correct." He said UFOs "exist. . . . We have not been able to duplicate their performance. . . . All we know is, we didn't make them, and it's pretty certain they didn't orig-

Robert I. Sarbacher in 1962.

inate on the earth." The issue was so sensitive that "it is classified two points higher even than the H-bomb. In fact it is the most highly classified subject in the U.S. government at the present time." Sarbacher refused to say more.

Smith, who died in 1961, mounted a small, short-lived UFO investigation, Project Magnet, for his government. Through official channels he tried unsuccessfully to learn more than Sarbacher's cryptic remarks had revealed. After the memo surfaced, ufologists found a listing for Sarbacher in *Who's Who in America,* citing his impressive scientific, business, and educational credentials.

Vannevar Bush in 1946.

When interviewed, Sarbacher said he had not personally participated in the UFO project, though he knew those who had, including RDB head Vannevar Bush, John von Neumann, and J. Robert Oppenheimer—three of America's top sci-

J. Robert Oppenheimer in 1964. Sarbacher identified him as one of the scientists involved in a top-secret UFO project that studied extraterrestrial hardware.

entists in the 1940s and 1950s. He had read documents related to the project and on occasion had been invited to participate in Air Force briefings.

"There were reports that instruments or people operating these machines were also of very light weight, sufficient to withstand the tremendous deceleration and acceleration associated with their machinery," Sarbacher wrote to an inquirer in 1983. "I remember in talking with some of the people at the office that I got the impression these 'aliens' were constructed like certain insects we have observed on Earth, wherein because of the low mass the inertial forces involved in operating of these instruments would be quite low. I still do not know why the high order of classification has been given and why the denial of the existence of these devices." Sarbacher could not recall where the crashes had taken place, but he did remember hearing of "extremely light and very tough" materials recovered from them.

Sarbacher's story never varied, and he resisted the temptation to elaborate or speculate. All who interviewed him were impressed. Still, his story could not be verified, since the persons he named were all dead. Sarbacher himself died in the summer of 1986.

ROSWELL UNRAVELED

The world was led to believe that the debris Mac Brazel had found near Corona, New Mexico, in 1947 was the remains of a weather balloon. For three decades, only those directly involved in the incident knew this was a lie. And in the early 1950s, when an enterprising reporter sought to reinvestigate the story, those who knew the truth were warned to tell him nothing.

The cover-up did not begin to unravel until the mid-1970s, when two individuals who had been in New Mexico in 1947 separately talked with investigator Stanton T. Friedman about what they had observed. One, an Albuquerque radio station employee, had witnessed the muzzling of a reporter and the shutting down of an in-progress teletyped news story about the incident. The other, an Army Air Force intelligence officer, had led the initial recovery operation. The officer, retired Major Jesse A. Marcel, stated flatly that the material was of unearthly origin.

The uncovering of the truth about the Roswell incident—so called because it was from Roswell Field, the nearest Air Force base, that the recovery operation was directed—was an excruciatingly difficult process. It continues to this day, even after publication of several books and massive documentation gleaned from interviews with several hundred persons as well as from other evidence. The Roswell incident is the most important case in UFO history. It has the potential, not to settle the issue of UFOs, but to identify them as extraterrestrial spacecraft. It is also the most fully investigated. The principal investigators have been Friedman, William L. Moore (coauthor of the first of the books, *The Roswell Incident* [1980]), Kevin D. Randle, and Donald R. Schmitt. Randle and Schmitt, associated with the Chicago-based Center for UFO Studies (CUFOS), authored the most comprehensive account so far, *UFO Crash at Roswell* (1991). Friedman's book, *Crash at Corona,* written with Don Berliner, was published in 1992. From this considerable research, the outlines of a complex, bizarre episode have emerged.

Eighth Air Force Commander Brigadier General Roger Ramey, acting under orders from General Clements McMullen at the Pentagon, concocted the weather balloon story to "put out the fire," in the words of retired Brigadier General Thomas DuBose, who in July 1947 was serving as adjutant to Ramey's staff. The actual material, all who saw it agreed, could not possibly have come from a balloon. For one thing, there was far too much of it. For another, it was not remotely like the wreckage of a balloon. Major Marcel described it:

> [We found] all kinds of stuff—small beams about ⅜ths or a half-inch square with some sort of hieroglyphics on them that nobody could decipher. These looked something like balsa wood and were of about the same weight, although flexible, and would not burn. There was a great deal of an unusual parchmentlike substance which was brown in color and extremely strong, and a great number of small pieces of a metal like tinfoil, except that it wasn't tinfoil. . . . [The parchment writing] had little numbers and symbols that we had to call hieroglyphics because I could not understand them. . . . They were pink and purple. They looked like they were painted on. These little numbers could not be broken, could not be burned . . . wouldn't even smoke.

The metallic material not only looked strange, it acted peculiarly. It had memory. No matter how it was twisted or balled up, it would return to its original shape, with no wrinkles. One woman who saw a rolled-up piece tossed onto a table watched in astonishment as it unfolded itself until it was as flat and wrinkle-free as the tabletop. When an acetylene torch was turned on samples of the material, they barely got warm and could safely be handled a moment or two later.

Air Force searchers scoured the recovery site until they had picked up what they thought were all pieces, however minuscule, of the crashed vehicle. Two years later, when Bill Brazel, Mac's son, let it be known he had found a few pieces the soldiers had missed, an Air Force officer called on him and demanded them. He handed them over without argument. Young Brazel knew how serious the military was about all this. After all, in July 1947 the Air Force had held his father incommunicado for days and made certain (through threats and, it is suspected, a large bribe) that he would never again talk about his discovery.

The material was secretly flown out of Eighth Army Headquarters in Fort Worth, Texas, to Wright Field (later Wright-Patterson AFB) in Dayton, Ohio. According to an officer who was there, Lieutenant Colonel Arthur Exon (who would become commander of the base in the mid-1960s), the material underwent analysis in the Air Force's material evaluation laboratories. Some of it, he recalled, was "very thin but awfully strong and couldn't be dented with heavy hammers. . . . It had [the scientists] pretty puzzled. . . . [T]he overall consensus was that the pieces were from space."

EXTRATERRESTRIAL BIOLOGICAL ENTITIES

But it wasn't just metal that had arrived at Wright Field.

Investigators have determined that a second crash—of a second, smaller, relatively more intact machine—occurred at a site some miles to the southeast of the Corona site, in the direction of Roswell. There military personnel found, amid the wreckage, four bodies. They were not the bodies of human beings.

This aspect of the Roswell story is the most fantastic, unbelievable, and difficult to document. The Air Force went to extraordinary lengths to hide it even from some of those who participated in the recovery of the material at the Corona site. Yet from the meticulous (and ongoing) research of Schmitt and Randle, we get the testimony of credible individuals who were involved, directly or indirectly, with the recovery of extraterrestrial remains. According to Exon, who heard the story from Wright personnel who had examined the bodies at the base, "they were all found . . . in fairly good condition."

Those who participated in the recovery of the bodies have provided consistent descriptions of what these "extraterrestrial biological entities" (the official designation, according to some unconfirmed accounts) looked like. They were four to five feet tall, humanoid, with big heads, large eyes, and slitlike mouths. They were thin and had long arms with four fingers. An Army nurse who worked on the initial autopsy at

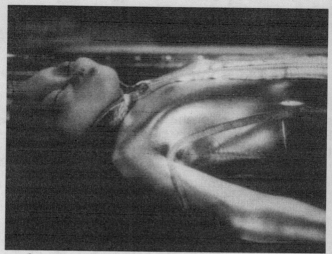

Impressionable people believe this widely published photograph shows the body of an extraterrestrial humanoid recovered from a UFO crash. In fact, the picture, taken in 1981, shows a wax doll in a Montreal museum.

Roswell remarked on how fragile the skull and bones were. Within hours the bodies were put into large, sealed wooden crates, loaded into the bomb pit of a B-29, and flown to Fort Worth Army Air Field. From there they went almost immediately to Wright Field.

Participants kept silent for years. Finally, as initial reports of the Roswell incident began to appear in the 1980s, they began to confide to close friends or family members what they had seen. Even then they were uneasy, still afraid of getting into trouble. One participant, Captain

Oliver ("Pappy") Henderson, flew the plane that first spotted the bodies. Apparently, judging from what he told his family, he also saw the bodies up close. Sergeant Melvin Brown rode in a truck with the bodies from the crash site to Roswell Field, then stood guard at the hangar where they were first stored.

Several persons who were at Wright Field or who knew individuals who were have testified to the arrival of wreckage and bodies at Wright in July 1947. One of these, retired General Exon, says a top-secret committee was formed to oversee the investigation of this and other highly classified UFO incidents. Nearly 20 years later, when he took command of the base, the committee was still operating. It had nothing to do with Project Blue Book, the poorly funded, inadequate project that apparently served little more than a public relations function. As Brigadier General Bolender had indicated in the internal Air Force memorandum quoted earlier (see Chapter 2, "UFOs: The Official Story"), UFO reports "which could affect national security . . . are not part of the Blue Book system."

Echoes of the Roswell incident have been heard for decades in popular folklore about secret rooms and buildings at Wright-Patterson AFB where government personnel study physical and biological proof of alien visitation. Most of these rumors—but not all—are "friend-of-a-friend" tales. Retired Wright-Patterson employee Norma Gardner claimed before her death ("Uncle Sam can't do anything to me once I'm in my grave") to have catalogued UFO material, including parts from the interior of a machine that had been brought to the base some years earlier. She also said she had typed autopsy reports on the bodies of occupants; once, moreover, she saw two of the bodies as they were being moved from one location to another. From her description—if she was telling the truth—she must have seen the Roswell entities. In the mid-1960s Senator Barry Goldwater, a brigadier general in the Air Force reserve, asked his friend General Curtis LeMay about the rumors. Goldwater told *The New Yorker* (April 25, 1988) that LeMay gave him "holy hell" and warned him never to bring up the subject again.

CASE STUDIES

SAUCER, SICKNESS, SECRECY

A UFO sighting on the evening of December 29, 1980, changed the lives of three Texans forever—and not for the good.

While driving through the southern tip of the east Texas piney woods, north of Houston, Betty Cash, Vickie Landrum, and Vickie's seven-year-old grandson, Colby, came upon a huge, diamond-shaped object just above the trees and about 130 feet away. Cash, who was driving, hit the brakes, and she and the elder Landrum stepped outside to see. Immediately they noticed intense heat. Their faces felt as if they were burning. When

Vickie Landrum reentered the car and touched the dashboard to steady herself, she left a handprint.

Blasting fire and heat, the UFO slowly ascended. Suddenly, numerous helicopters—23 in all—appeared from all directions, positioning themselves near the strange craft. By this time the witnesses were back in the car and watching the spectacle from their moving vehicle. (Other motorists saw the object and the helicopters from different, more distant locations.) Eventually, the flying objects were lost to view. Unfortunately the episode was only beginning.

Back home the three fell sick, Cash most severely. She suffered blisters, nausea, headaches, diarrhea, loss of hair, and reddening of the eyes. On January 3, unable to walk and nearly unconscious, she was admitted to a Houston hospital. Vickie and Colby were experiencing the same symptoms, though less severely.

The witnesses' health problems have continued to this day. In September 1991 Cash's personal physician, Dr. Brian McClelland, told the Houston *Post* that her condition was a "textbook case" of radiation poisoning, comparable to being "three to five miles from the epicenter of Hiroshima." For years the three have pursued their case through the courts, seeking answers and redress, but official agencies deny any knowledge of the incident—even though the helicopters have been identified as twin-rotor Boeing CH-47 Chinooks, which are used by both the Army and the Marines.

MAJESTIC MYSTERY

In December 1984 a package with no return address and an Albuquerque postmark arrived in UFO investigator Jaime Shandera's mail in North Hollywood, California. Inside was a roll of 35mm film. When developed, it turned out to contain eight pages of an alleged briefing paper, dated November 18, 1952, in which Vice Admiral Roscoe Hillenkoetter told President-elect Dwight Eisenhower of the recovery of the remains of two crashed spaceships. In the first of these crashes, in early July 1947, authorities recovered the bodies of four humanoid beings. According to the document, which appended a copy of what was supposed to be the actual executive order, President Harry Truman authorized the creation of a supersecret group called "Majestic 12" (MJ-12 for short) to study the remains.

Acting on a tip from sources who claimed to represent Air Force intelligence, Shandera and his associate William Moore (coauthor of *The Roswell Incident*) flew to Washington, D.C. They searched the National Archives looking for references in official documents to MJ-12. They found a July 1954 memo from General Robert Cutler, an Eisenhower assistant, referring to an "MJ-12 SSP [Special Studies Project]" to be held at the White House on the 16th of that month.

In the spring of 1987 an unknown individual, allegedly associated with an intelligence agency, gave British writer Timothy Good a copy of the MJ-12 document. Upon learning Good was going to disclose its existence to the press, Moore and

Shandera released their copy, along with the Cutler memo. The result was a massive uproar, including coverage in *The New York Times* and on *Nightline,* an FBI investigation, and furious controversy that continues to this day.

For various technical reasons most investigators agree that the MJ-12 document is a forgery, but the identity of the forger remains a deep mystery that even the FBI cannot crack. The forger apparently had access to obscure official information, much of it not even in the public record, leading to suspicions that an intelligence agency created the document for disinformation purposes. Whatever the answer, the MJ-12 document is surely the most puzzling hoax in UFO history.

SHIPS IN THE FOREST

We may never know the full story of what happened between December 26 and 27, 1980, in Rendlesham forest, on England's east coast between two U.S. Air Force bases, Woodbridge and Bentwaters. The incidents remain shrouded in secrecy. What we do know—learned through a painstaking, years-long investigation by civilian researchers—is fantastic enough.

Just after midnight on December 26, eyewitnesses and radar screens followed an unidentified object as it vanished into the forest. Soldiers dispatched to the site encountered a luminous, triangular-shaped craft, ten feet across and eight feet high, with three legs. The UFO then retracted the legs and easily maneuvered its way through the trees. The soldiers chased it into a field, where it abruptly shot upward, shining brilliant lights down on them. At that moment the witnesses lost consciousness. When they came to, they were back in the forest. Other troops sent to rescue them found tripod landing marks where the object apparently had rested.

The following evening observers reported seeing strange lights. The deputy base commander, Lieutenant Colonel Charles Halt, led a larger party into Rendlesham. There Halt measured abnormal amounts of radiation at the original landing site. Another, smaller group, off on a separate trek through the forest, spotted a dancing red light inside an eerily pulsating "fog." They alerted Halt's group, who suddenly saw the light heading toward them, spewing forth a rainbow waterfall of colors. Meanwhile, the second group now watched a glowing domed object in which they could see the shadows of figures moving about. During the next hour both groups observed these and other darting lights.

Cable News Network learned that films and photographs were taken of these events, despite official denials. According to curiously persistent rumors, never verified but never conclusively disproved, occupants were encountered at some point during the event.

When a constituent told him about the incident, U.S. Senator James Exon launched an extensive but secret inquiry. He has never revealed his findings. He says only that he learned "additional information" that ties the Rendlesham case to "other unexplained UFO incidents."

STRANGE SHAPES OVER BELGIUM

On his way to pick up his wife, a Belgian man passed through an industrial park outside Bierset, Belgium, at 6:45 on the evening of November 29, 1989. In the park he was startled to see a large, lozenge-shaped structure hovering near the highway. It was at an altitude of slightly more than 300 feet and was about 500 feet away from him.

Slowing his car, the driver rolled down the window and observed the UFO carefully. The body consisted of what looked like a dark metal. Lights flashed along the side, and three searchlight beams—one red, one green, and one white—swept the ground. All the while a soft, throbbing sound emanated from the vehicle, which appeared to be bigger than a Boeing 707. It still hovered as the witness resumed his journey.

For much of 1989 and 1990, Belgium—where most of the action took place—and such neighboring countries as France and Germany hosted comparable sightings of incredible aerial phenomena. The objects were often, though not always, described by their witnesses as triangular in shape and enormous in size. Some of the sightings were confirmed by ground and air radar; some were recorded on film and video. They received international attention. They were the subject of intensive investigation by an alliance of military officials, scientists, and civilian ufologists. In the end they concluded that the UFOs were the product of an advanced technology, source unknown.

"EXPLAINING" UFOs

At least three out of every four UFO reports turn out to have conventional explanations. Typical IFO (identified flying object) sightings are of stars, planets, meteors, balloons, advertising planes, and optical illusions, or are hoaxes. Skeptics argue that the other 25 percent of reports could probably be explained if additional information were available. This argument sounds logical but is in fact demonstrably false. Between 1952 and 1955 the Battelle Memorial Institute in Columbus, Ohio, a think tank that does classified analytical work for the U.S. government, studied Project Blue Book's collection of UFO reports. The Institute established, with virtual statistical certainty, that the unexplained sightings were fundamentally different from both explained sightings and those sightings with insufficient information for evaluation. The "unknowns" came from the best-qualified observers, the sightings were of longer duration than the "knowns," and the unknown objects seldom bore any resemblance in appearance or behavior to their conventional counterparts. All in all, the unknown category consisted of those reports in which the *most* information was available—precisely the opposite of what would be expected had the reports been potentially explainable.

HOW SECRET IT IS

The late comedian Jackie Gleason's second wife, Beverly, tells a strange story that she swears is true. One evening in 1973, she wrote in an unpublished book on their marriage, Gleason returned to her Florida home badly shaken. After first refusing to tell her why he was so upset, Gleason confided that earlier in the day his friend President Richard Nixon had arranged for him to visit Homestead Air Force Base in Florida. Upon his arrival armed guards took Gleason to a building at a remote location on the site. There, Gleason, who harbored an intense interest in UFOs, saw the embalmed bodies of four alien beings, two feet long, with small bald heads and big ears. He was told nothing about the circumstances of their recovery. He swore his wife to secrecy, but after their divorce Beverly freely discussed the story. In the mid-1980s, when ufologist Larry Bryant sued the U.S. government in an attempt to get it to reveal its UFO secrets, he tried without success to subpoena Gleason to testify. Gleason never commented on Beverly's report.

U.S. GOVERNMENT IN CONTACT WITH ETs?

Former CIA operative Victor Marchetti, coauthor of the best-selling *The CIA and the Cult of Intelligence* (1974), thinks the U.S. government maintains secret contacts with extraterrestrials. He bases his suspicion—he admits he cannot prove it—on stories he heard while working at "high levels of the CIA." These tales alleged that the National Security Agency (NSA), which collects electronic intelligence, had received "strange signals," said by intelligence sources to be of extraterrestrial origin. Marchetti could learn nothing about the content of these communications, which had a level of secrecy that was extraordinary even by the standards of the supersecret NSA. In the 1980s UFO investigators William Moore and Jaime Shandera heard comparable tales from Air Force intelligence sources. No good evidence backs up these tales, but they are undeniably intriguing, if only because of who is telling the tales.

Victor Marchetti has written that official efforts to discount UFOs "have all the earmarkings of a classic intelligence cover-up."

BEYOND THE ULTIMATE SECRET

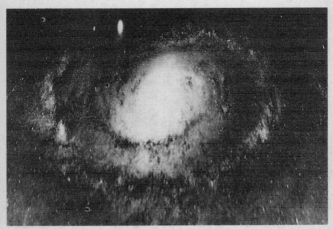

If life is common in the universe, as many astronomers believe, visitation from other worlds is not only possible but probable. The UFO phenomenon, which has resisted nearly five decades' efforts to explain it in conventional terms, may represent the first solid evidence that we are not alone.

On the evening of October 25, 1990, a Lancaster, California, woman was watching a *Hard Copy* television show devoted to a remarkable series of events that had taken place in Belgium over the past year. Beginning in November 1989, many Belgians, including a number of police officers, had reported seeing a huge, triangle-shaped UFO with three brilliant lights. When the object appeared late at night on March 30, 1990, witnesses had called the air force radar station at Glons. The operators there had picked up the object on their screen and notified colleagues at the station at Semmerzake. The UFO had showed up on their screens, too.

Over the next 55 minutes, radar experts had determined that there was no prosaic explanation, such as false echoes caused by temperature inversions, for the image. Two F-16s were sent out to intercept the object, which had showed up on the radars of both planes. It looked, according to a subsequent account in the magazine *Paris-Match,* like a "little bee dancing on the scope." Six seconds after the jets achieved a radar lock-on, the object *in one second* accelerated from 175 miles per hour to 1,100 miles per hour, descending 3,000 feet in the same amount of time.

During the *Hard Copy* program, the California woman happened to glance outside her window. Even as she watched the show, she could see through her window a strange, triangle-shaped craft maneuvering in the northern sky. Visible for the next ten minutes, it carried a bright light at each of its three corners. Sightings of triangular objects go back to at least the 1950s. Over the past two decades, reports of such UFOs have increased markedly and have been logged all over the world.

On the evening of August 25, 1951, Hugh Young, a security guard at Sandia Base, a sensitive atomic installation near Albuquerque, New Mexico, was off duty. He and his wife, Emily, observed a bizarre sight: a huge flying wing one and a half times the size of a B-36 wing span. It came in from the north, moving silently at less than 1,000 feet and at approximately 300 miles per hour, and passed over the witnesses' trailer home. The Youngs told Air Force investigators that the object had dark bands running from front to back; along its trailing edges were six to eight pairs of glowing lights.

In the late summer of 1951, sightings of similar "flying wings" occurred frequently in the Southwest. One was photographed on August 30 as it passed over a Lubbock, Texas, neighborhood just before midnight. UFOs of this description first appeared in early July 1947, during the wave that followed Kenneth Arnold's classic report of "flying saucers" on June 24. In one instance, on July 6, a Darlington, South Carolina, attorney reported seeing ten or 12 of them flying in—fittingly—V formation.

In 1983 and 1984 thousands of persons in seven densely populated counties in New York and Connecticut observed enormous structured objects. The UFOs were described as giant boomerangs or Vs with lights on the side. In some instances the structures had bright searchlights sweeping down from the craft's underside. Sometimes they hovered no more than ten or 20 feet off the ground, and sometimes they moved— so joggers testified—at walking speed. They were also capable of attaining astonishing speeds in seconds.

Numerous sightings of such boomerangs have occurred from the mid-1980s to the present in Antelope Valley, California, which encompasses Lancaster, Rosamond, and Palmdale. Locals have reported, besides boomerangs and triangles, fast-moving discs. In one especially spectacular instance, which took place on October 26, 1988, a valley couple observed a slow-moving, immense boomerang—over 600 feet in span, they estimated—which soon was joined by an identical structure. Behind the second stretched a formation of nearly 20 smaller, disc-shaped craft. Other Antelope Valley residents attest that the triangles and boomerangs move at every speed from walking to ultrasupersonic speed, sometimes passing from one to the other in seconds.

THE RIDDLE INSIDE THE ENIGMA

According to persistent reports from individuals who appear to be sane, sincere, and well placed, the existence of these and other sorts of strange craft constitute some of the deepest secrets of the U.S. government. Information concerning them is confined to a tiny group of officially connected individuals with a need to know. Few elected officials are privy to these secrets because elected officials are notorious for leaking sensitive information if it serves their political purposes.

Nonetheless, over the years a number of individuals claim to have seen such craft in official custody. They have confided their stories to journalists and investigators. Their stories, while unproven, are remarkably consistent and would seem credible were it not for their fantastic content.

Those who are convinced of the authenticity of such sightings argue that the witnesses are sober and reliable. Some, in fact, are trained observers—scientists, engineers, pilots—unlikely to be mistaken about what they have seen. Moreover, photographs, films, and videotapes show unknown objects exhibiting extraordinary performance characteristics far beyond the capabilities of known aircraft. Some individuals who report getting particularly revealing film, photo, or video evidence say that their pictures were confiscated by threatening individuals flashing official credentials.

Government and military spokespersons emphatically deny that such craft exist. If they did, Air Force Secretary Donald B. Rice wrote in a letter published in the *Washington Post* (December 27, 1992), "I'd know about it—and I don't. . . . Some reported 'sightings' will probably never be explained simply because there isn't enough information to investigate. Other accounts . . . are easily explained, and we have done so numerous times on the record."

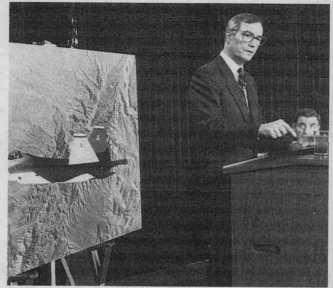

Secretary of the Air Force Donald B. Rice in 1991

Panels of scientists convened to assess the evidence have made short work of it. A *New York Times* article of January 29, 1993, characterized the sightings as mere "yarns." John E. Pike of the Federation of American Scientists says such imagined events are psychosocial in origin, the product of "some of the deeper anxieties of contemporary American society."

Those who believe the reports respond that the debunkers do not know what they are talking about. They have not interviewed the witnesses or gone out into the field to collect evidence or attempted a serious, open-minded analysis of the available data.

Meanwhile the reports continue, and the debate goes on. This debate, by the way, is not about reports of UFOs. It concerns recent alleged developments in aviation technology. However, the language traditionally employed against UFO sightings is used to dismiss reports of these extraordinary aircraft. As Pike remarks, tellingly if ungrammatically, an "understanding of the mystery aircraft phenomena [sic] is impossible outside the context of the UFO phenomena [sic]." He means that as a putdown, but his words ring true in all sorts of ways he probably does not suspect.

MYSTERIES AT AREA 51

Area 51, also called Dreamland, is located at a corner of the Nevada Test Site, where highly classified national defense projects have been conducted for decades. Such spy planes as the U-2 and the SR-71 were developed there, as were the Stealth aircraft and the technology associated with the Strategic Defense Initiative ("Star Wars").

In 1984 a curious, unexplained item appeared on the defense budget next to the SR-71. (The SR-71 was taken out of service in early 1990, supposedly because spy satellites could now perform the same reconnaissance functions.) It said simply, "Aurora." The indication was that Aurora, whatever it was, was being built, as was the SR-71, by the Lockheed Corporation, with Rockwell International's Rocketdyne division responsible for the engines.

The Lockheed plant in Burbank, California

The following year residents of southern California and other areas, including northern Europe, began hearing earthquakelike rumbles and seeing extraordinary flying objects moving at several times the speed of sound. The craft looked like giant triangles. By 1990 reporters for *Aviation Week & Space Technology,* the leading magazine in its field, had become convinced that a "quantum leap in aviation" had taken place under conditions of great secrecy. In the October 1 issue John D. Morrocco, an *Aviation Week &*

Space Technology editor, noted sightings of triangles "over the northern end of the Antelope Valley, near Edwards AFB and Mojave, California, as well as in central Nevada."

Morrocco and other reporters, notably *Aviation Week & Space Technology* colleague William B. Scott, also collected sightings of boomerang- or wing-shaped craft, as well as testimony from informants who claimed to have seen or even worked on such vehicles. Other witnesses recounted observations of diamond-shaped craft and fast-moving nocturnal lights.

As these stories started attracting widespread attention, denials and controversy erupted. In time the Federation of American Scientists endorsed the government's claims that no Aurora project or extraordinary aircraft existed outside witnesses' imaginations. Nonetheless the evidence was there to anyone willing to take Nevada back roads to the outskirts of Area 51, where almost nightly mysterious lights streaked across the sky.

These objects are indistinguishable from what are ordinarily thought of as UFOs, whose existence officialdom and the scientific establishment also deny. Their similarity to UFOs probably lies at the root of the nervousness with which the scientific establishment greets the Aurora matter. After all, if witnesses are accurate when they report these extraordinary terrestrial aircraft, how can they say witnesses are not accurate when they report extraordinary extraterrestrial spacecraft?

But that is just the beginning of it.

According to virtually all accounts, prototypes of the Aurora aircraft were test-flown no earlier than the late 1960s. Yet sightings of eerily similar objects, described as UFOs, go back three to four decades. All but the southern California sightings cited earlier are considered "UFO" reports. What in the world—or elsewhere—is going on here?

Tales of secret projects linked to UFO technology have circulated for years. Some are traceable to informants who seem credible and well placed. A *Las Vegas Review-Journal* article mentions allegations that the "remains of an alien spacecraft" are stored at the Nevada Test Site. Aviation journalist James C. Goodall, who has investigated such stories, writes, "Rumor has it that some of these systems involved force field technology, gravity drive systems, and 'flying saucer' designs. Rumor further has it that these designs are not necessarily of Earth human origin, but of who might have designed them or helped us do it, there is less talk."

Perhaps the wildest element of all shows up in two UFO-abduction claims from Antelope Valley. In each case the abductee claimed, under hypnosis, to have been snatched up by the standard little gray humanoids. But in these reports, the humanoids were accompanied by men in Air Force uniforms. Though ufologist William Hamilton, who has investigated these reports, takes them seriously, caution and skepticism seem the wisest response.

Tales of crashed-saucer projects may be unverified, but they are nonetheless remarkably persistent. If the Roswell incident occurred, as a huge body of testimony attests it did, it follows that research projects—extremely sensitive, deeply classified ones—would have been mounted in an effort to crack the secrets of extraterrestrial technology. Much of this sort of research would have been done at the Nevada Test Site. In the absence of solid evidence, of course, such notions must remain purely speculative. Rumors, after all, must never be confused with evidence.

Still, one can only wonder why the Aurora aircraft (assuming they exist) look and behave so much like UFOs. Can this be no more than coincidence—especially when the similarity is so great that often it seems impossible to tell which is which?

When Morrocco's *Aviation Week* article recounted sightings in the American Southwest, a National Public Radio reporter asked him if there might be some connection between those and the Belgian reports. Morrocco was hard pressed to explain the difference between the Southwestern reports and what he called "mere sighting of UFOs." Privately *Aviation Week* reporters express great interest in UFO sightings.

At the moment five interpretations seem hypothetically possible: (1) All triangles and boomerangs are secret Aurora aircraft. (2) All such objects, wherever observed, are UFOs. (3) Some triangles and boomerangs are UFOs, and some are Aurora aircraft. (4) It is mere happenstance that the secret aircraft and the UFOs are identical in appearance. (5) The rumors are true. It is no coincidence that the UFOs and the Aurora aircraft look the same.

At the moment this is an intriguing mystery with potentially staggering implications. Wherever it goes from here, we may be certain that it will be *the* UFO story of the 1990s.

CASE STUDIES

ENIGMATIC ENCOUNTER OVER PUERTO RICO

Since the late 1980s, Puerto Ricans have reported everything from balls of light to little gray beings to mysterious aircraft. These events, almost entirely ignored in the mainland United States, have received extensive coverage in the local press.

Perhaps the most disturbing of all these bizarre stories is said to have occurred over the southwestern region of the commonwealth on the evening of December 28, 1988. At 7:45 residents of Betances heard jet planes roaring through the western sky. "My wife went out to see them," one witness, Wilson Sosa, recalled, "and in a few minutes came back calling for me to come look. I went out and there was a huge UFO blinking with many colored lights, coming over the Sierra Bermeja. I ran to get my binoculars—I always keep them handy, since we see so many strange lights here—and I could clearly see it was a big

triangle with a little curve at the rear side. It came
forward and turned, then came lower. Then we
saw two jet fighters behind it. . . . It was bigger
than the baseball park here, and it just hung
there. There was a jet at the left rear of the craft,
and one at the right. They looked like mosquitoes
compared to the UFO, it was so big.

"Then the people all hollered because it looked
like the jet at the rear was going to collide with
the craft. . . . Then the plane disappeared. I don't
know if it went into the UFO or what happened,
but its lights were just gone. And then the UFO
turned toward the west, and the second jet disappeared, too. Its lights were gone, and the engine
sound was gone."

The UFO stayed for a while longer. It descended and hovered, then separated into two pieces
from apex to base of the triangle, showering red
sparks. The two smaller triangles departed rapidly in different directions.

Puerto Rico is riddled with military bases and
aircraft carriers from which the jets—identified
as F-14s—could have come, but authorities
denied that any military aircraft were missing. No
aircraft disaster or pilot losses can be traced to
the period in question.

Nonetheless 113 witnesses swore to the event,
and sources at a local U.S. Navy base told of radar
records of the incident. Witness Sosa claimed that
soon after the sighting, two Air Force officers visited him and warned him not to discuss the incident—a warning he pointedly ignored, without
dire consequences.

What happened? There seem only two possibilities: mass hallucination or an extraordinarily successful cover-up of an encounter with decidedly unsettling implications.

INTIMATIONS OF INFINITY

If intelligent life exists elsewhere, according to a view held by many astronomers, it is likely to look generally like us. It is also probable, according to Carl Sagan, Frank Drake, and other specialists in exobiology, that the galaxy teems with extraterrestrial civilizations. Some of these civilizations are older than Earth's and have superior technologies. "It is extremely probable," Michael D. Swords of Western Michigan University writes in a survey of the scientific literature on the subject, "that some, if not all, of these advanced civilizations have the means, albeit with difficulty, of traversing interstellar space. And it is essentially a certainty that these advanced life-forms have several instincts/motivators/behaviors in common with *Homo sapiens,* one of which (curiosity) may be particularly germane to such journeys."

Visitation from elsewhere is not just possible, it's something that we should expect. From every indication such visitation is happening now. It remains a secret only to those who have not seen it in operation with their own eyes, read the UFO evidence with an open mind, or examined—at some secure facility somewhere—the metal and the bodies that fell out of the sky one night in 1947 and forever proved we are not alone.

INDEX

INDEX